The Technical Writing Process

The Technical Writing Process

Marilyn Schauer Samuels

Associate Professor of English,
Director of Technical and Professional Writing
Case Western Reserve University

New York Oxford
OXFORD UNIVERSITY PRESS
1989

Oxford University Press

Oxford   New York   Toronto
Delhi   Bombay   Calcutta   Madras   Karachi
Petaling Jaya   Singapore   Hong Kong   Tokyo
Nairobi   Dar es Salaam   Cape Town
Melbourne   Auckland

and associated companies in
Berlin   Ibadan

Library of Congress Cataloging-in-Publication Data

Samuels, Marilyn Schauer.
The technical writing. process.

Includes index.
1. Technical writing.   I. Title.
T11.S26   1989   808′.0666   87-7834
ISBN 0-19-503679-4

2 4 6 8 9 7 5 3 1
Printed in the United States of America
on acid-free paper

To my beloved son,
David Joseph

To the Instructor

This textbook aims to assist you in making contemporary research on writing and rhetoric, cognitive science, sociology of knowledge, ethnography, and collaborative learning available to your students for practical use in technical writing.

Through a reader-centered process approach, the chapters provide explanations, examples, case studies, and exercises for technical writing students of all disciplines.

The text is divided into three sections: contexts and attitudes, processes, and applications. Because a recursive learning process is the system within which this text is designed to operate, these sections are not mutually exclusive. The ordering of the sections and chapters is designed for students to be continually recalling old knowledge, acquiring new understanding, and experiencing an ongoing dialogue between old, recent, and new information and its usability as they move through the readings and assignments. By presenting the concept and example of a process model of writing, the material enables students to develop individual writing processes adaptable to each of their communication needs.

SECTION I: CONTEXTS AND ATTITUDES

Chapter 1, "Technical Reading Processes," introduces students to the technical reader as an individual and as part of a reading community. Understanding how readers read, from physiological, psychological, and sociological perspectives, is presented as a first key to effective communication.

Chapter 2, "Technical Writing as Salesmanship," and chapter 3, "Patterns of Logic," give students two different perspectives on creating an interactive relationship with their readers.

SECTION II: BEFORE, DURING, AND AFTER

Chapter 4, "Before You Write," explores with students the problem-solving and decision-making processes that precede and facilitate effective writing.

Chapter 5, "As You Write," uses case studies and examples to take students through the process of composing and demonstrate problems and solutions that occur in process.

Chapter 6, "Using Visual Aids," presents illustration as an integral part of the writing process. It familiarizes students with the types of visual aids, the strengths and weaknesses of each, and guidelines for deciding what to use when.

Chapter 7, "After You Write," examines revision processes both before and after the document has been distributed to readers.

SECTION III: THE PROCESS APPLIED

Chapters 8 through 14 focus students on adapting the information acquired in chapters 1 through 7 to specific types of writing and of communication situations. By studying and discussing actual examples and contexts, students learn to write job applications (chapter 8), letters, memos, and minutes (chapter 9), descriptions and instructions (chapter 10), proposals (chapter 11), feasibility and progress reports (chapter 12), long reports such as environmental impact statements and bids for government contracts (chapter 13), and to prepare and deliver oral reports (chapter 14).

APPENDICES

Four long reports are provided in the appendices for use as suggested throughout the text and as material for creating your own lessons and exercises: a guide to *Cost Control Procedures in the Restaurant Business,* a *NASA* technical note on a *Towed Parawing Glider Model;* a guide to the *DECSYSTEM 20;* and excerpts from the Solar Energy section of an environmental impact statement (EIS), *The Alternative Fuels Demonstration Program.*

AUDIENCE AND ADAPTATIONS

The audience for this book is primarily junior- and senior-level college students. The fourteen chapters are designed for a 15-week semester or for a 10-week quarter when the first section is covered in one or two meetings instead of three and/or some of the application chapters in section III are combined or omitted depending on the needs of the group.

The author has used this text successfully in freshman and sophomore composition courses, in-house seminars, review courses for graduate students, and training courses for graduate teaching assistants. Slight adaptations of order and style of presentation and additions or deletions of examples make this book a valuable resource for almost any learning situation in which the subject is the technical writing process specifically or professional communication skills in general.

The text may also be adapted to different teaching styles. By changing the order in which chapters are read, or by placing more focus on some chapters than on others, or by substituting your own samples or using the book's examples for different purposes, you may choose to emphasize pre-writing, revision, style, problem-solving, or document design.

The use of small group workshops is illustrated and recommended throughout. In twenty years of teaching (at CCNY with the late Mina Shaughnessy, at Case Western Reserve University, and in corporate classrooms), I have found no better way of explaining writing skills than the mutual and individual discovery that takes place when writers are confronted with their readers and when writers collaborate as group readers to discuss their own or professional examples.

If you discover other interesting adaptations of the book's materials or would like to share suggestions for additions or changes, I would welcome hearing from you.

To the Student

This book is an example of a selling job. It is meant to sell you on the idea that effective technical communication is a vital, exciting part of your profession, as well as an absolutely necessary and attainable skill.

It will teach you how to add writing to the well-managed subordinate skills that enable your success in your chosen profession, or it will give you the foundation for pursuing a career in the profession of technical communication.

You will learn to use words instead of being intimidated by them; you will promote successful technology transfer as well as improve your own professional opportunities.

As with all new skills, you will need to become acutely conscious of how writing and reading are done before you can take the new communication skills you will be acquiring for granted. This process will not be painless, but the effort will pay off.

You will be developing recyclable approaches to writing that you can adapt to any writing or speaking situation requiring you to convey information or ideas to others.

The guidelines and examples will assist you in developing your own effective communication processes and collecting your own successful papers as models. For example, the instructions for a game that you write as an assignment in Chapter 10 may serve as a model when you need to describe how to run an accounting program or how to monitor the electrons in a prosthesis. You may redesign and add to your samples and procedures as you advance in your career, until you have at least one good example in your files of every type of writing you do.

First, try to let go a little of any ideas you have acquired in the past about whether you ever can write, or ever should have to write. Keep an open mind, then let *The Technical Writing Process* help you to acquire a lifelong ability to communicate well. Many former students of mine have written to say that it works.

Contents

The Process: Contexts and Attitudes

CHAPTER 1

Technical Reading Processes: Cognition, Community, and Collaborative Learning

Currently, there are two schools of thought on how readers either derive or create meaning from their interaction with a text. One, the cognitive approach, focuses on the physiological and mental processes that take place inside the individual brain when confronted with a communication. The other, the sociology of knowledge school, focuses on environmental influences—social, political, cultural, ethnic, and professional—that affect how groups of readers representing a specialized community of inquiry or community of discourse will make meaning from written documents.

The first approach stresses mental processing patterns that all readers experience individually and attempts to discover and analyze the possible variations of these common patterns and their impact on the structuring of a text. The second approach concentrates on learned paradigms, tacit understandings, and other group mythologies which readers in a particular field of learning or in a particular industry or company may have in common. This group reader approach proposes that the commonly accepted beliefs, past knowledge, and expectations of any group of readers represents the standard against which each member of the group measures new communications in order to process and understand new information.

Reading behaviors based on cognitive processes may be described as "conversations" that readers engage in internally while moving through a text. The topics of these conversations include how the information in the text corresponds directly or by analogy with what the reader already knows, and what expected patterns in the text, such as comparison, highlighting, subordination, may be pursued in order to comprehend important points and the ways in which they are connected.

Reading behavior based on the sociology of knowledge and the concept of collaborative or group learning, according to Kenneth Bruffee, claims that individuals learn how to conduct these private conversations and thereby derive a text's meaning by first learning the conventions of conversing in groups (1). Knowledge is created by group conversations that lead to consensus. Accumulated group consensuses are internalized and continue to determine how group members process texts individually. In *Corporate Cultures,* for example, Deal and Kennedy demonstrate that what employees believe to be true of the culture or "myths" of their company will determine how they interpret each new company directive, memo, or report (2).

Reader-processing depends *both* on individual mental processes and their possible variations *and* on the determination of community culture and consensus. Therefore, as a technical writer, you can use your understanding of internal and social influences to communicate effectively.

INDIVIDUAL COGNITION

Begin by becoming aware of your own reading process. How often have you read several pages of a textbook or several paragraphs of a magazine or journal article without getting any sense of why the facts presented are being given to you—without seeing any connection between one point and the next? How often have you realized suddenly that you are about 20 pages into a book and have little sense that you have been reading it at all?

Do you remember what you did when these or similar experiences occurred while you were reading? Chances are that if the material was something other than required reading, you gave up on it, fell asleep, or picked up something else. If it was required reading, however, you probably

- discovered or else imposed a format by linking kinds or degrees of things; by making comparisons; by relating the new information to something you already know; by numbering or rearranging items, and so on;
- turned to the last page or chapter to find a concluding summary that would give you the main points or provide a context;
- skimmed the article, the index, or the table of contents to get an overview;
- read through the headings, chapter titles, or underlined sentences to get the highlights or a sense of the whole.

Communication models from the fields of electronics and cognitive psychology confirm that if technical writers do not structure documents to facilitate these problem-solving processes, readers will make their own meaning. If technical writers want readers to understand *their* meaning, they must create a structure that makes that meaning accessible. Otherwise, readers will impose a structure—not necessarily the one the writer intended.

In her book *Writing as Problem-Solving,* Linda Flower demonstrates the writer/reader interaction essential to communication transfer using an IEEE diagram of teletransmission. The writer (encoder) sends the message, and the reader (decoder) receives it. Meaning is transmitted only if both acts, encoding and decoding, are completed successfully. Creative readers, observes Flower, "remember not what we tell them, but what they tell themselves" while processing our text (3). For example, experiments have shown that:

- *Readers usually try to fit new information into a familiar framework.* If you've devised a new method for building centerboard sailboats, the reader will relate your report to what she/he already knows about boat construction or construction in general.

- *As readers read, they create a context or set of expectations about what is to follow.* If you have stated your intention to explain the difference between A and B, they will not expect you to prove that B is better than A. If you compare A and B without explaining why, they will draw their own conclusions about whether they are reading exposition or argument.

- *Readers sort information into categories or chunks in order to understand it.* If you have not divided your subject into clearly distinct parts (problems and solutions, soluble versus insoluble instances, and so forth), readers will discover a system of categorization or create their own.

Other observable traits of the individual reader are reviewed by Thomas N. Huckin in "A Cognitive Approach to Readability" (4). For example, the *schema theory* demonstrates that experts read differently than novices. Therefore, they need clues that call forth from their long-term memory the experience and knowledge necessary to understand the information.

The levels effect and the leading-edge strategy show that readers assume information underlined or otherwise highlighted is more important than other information in the text. Readers also make hierarchical distinctions in interpreting texts: something mentioned first is remembered better than something mentioned fifth. On the other hand, readers remember best what they have read most recently. That is why repeating main points in a conclusion is so important.

It is also important to know the different styles of reading. A. K. Pugh in *Silent Reading* (5) has identified five:

skimming	reading for the general drift of the passage
scanning	reading quickly for the purpose of finding specific items of information
search reading	scanning with attention to the meaning of specific items
receptive reading	reading for thorough comprehension
critical reading	reading for evaluation

If writers know which style their readers will use, they can design the text accordingly. If they suspect that more than one style will be used by each or some of their readers, they can cue and highlight the text for more than one anticipated pattern. In a recent study of reading patterns in on-line documentation, for instance, a software program called DOCUMENT kept transcripts when anyone accessed a document. It was found that most users looked up specific facts using an index of topics. Rarely was the full text or even one entire section retrieved. Having learned that readers mainly consult the reference or the error index, on-line documentation writers re-designed these indexes to facilitate quick, accurate problem-solving (6).

COMMUNITY CONSENSUS

Consciousness of readers' cognitive processes will improve your writing. But you also need to be aware of how your reader perceives as a member of a professional community—nuclear scientists, or classical musicians, for example—and as a member of a particular company or institution such as Pennsylvania Electric or the New York Philharmonic. "There is no single way of reading," declares critic Stanley Fish; "only ways of reading that are extensions of community perspectives" (7).

For example, Thomas Kuhn demonstrates in *The Structure of Scientific Revolutions* (8) that the theories accepted into a scientific community are those judged to be practical and acceptable within the group's current consensus on the criteria for truth. Scientific revolution occurs when it is the consensus of the majority that a new approach answers more questions and presents fewer problems than previous assumptions. This new approach then becomes the consensus, and all fresh developments will be judged on the basis of whether or not they serve the purposes of the new group-supported version of reality.

In other words, knowing the beliefs, the ways of looking at things that your reader holds in common with his/her professional community, assists you in structuring a text your reader can understand. Sometimes, this goal can be accomplished by reading company style manuals and company policy statements or philosophies. More often, you will need to interact formally and informally with the group you will be addressing individually or collectively in your writing.

Throughout the following chapters, methods of group interaction or collaborative learning for the purposes of writing and reading technical communications are demonstrated in the sample case studies. The exercises give you opportunities to practice collaborative learning by serving as test readers of each other's writing and coming to consensus on the group's individual observations of how different readers respond to the same text.

Combining your knowledge of cognitive reading processes and social/environmental influences on your readers, you will be able to communicate information to any reader, and to adapt the same information to different readers in different contexts.

REFERENCES

1. Kenneth Bruffee. "Collaborative Learning and the 'Conversation of Mankind.'" *College English* 46 (1984): 635–652.
2. Terrence Deal and Allan A. Kennedy. *Corporate Cultures* (Reading, Mass.: Addison-Wesley Publishing Company, 1982).
3. Linda Flower. *Problem-Solving Strategies for Writing* (New York: Harcourt, Brace, Jovanovich, 1981), 130.

4. Thomas N. Huckin. "A Cognitive Approach to Readability," in *New Essays in Technical and Scientific Communication,* ed. Paul V. Anderson, Carolyn R. Miller, and John Brockman (New York: Baywood Press, 1985), 90–108.

5. A. K. Pugh. *Silent Reading* (London: Heinemann, 1978).

6. T. R. Grill, Clement H. Luk, and Sally Norton. "Reading Patterns in Online Documentation . . . ," *Proceedings* of the 34th International Technical Communication Conference, pp. RET 111–114. Washington, D.C.: Society for Technical Communication, 1987.

7. Stanley Fish. *Is There a Text in This Class?: The Authority of Interpretive Communities* (Cambridge: Harvard University Press, 1980), 15.

8. Thomas E. Kuhn. *The Structure of Scientific Revolutions* (Chicago: University of Chicago Press, 1970).

CHAPTER 2

Technical Writing as Salesmanship

The technical writing process applies the principles of good salesmanship to communication: 1. Know the product/subject; 2. Know the customer/reader and his/her professional community; 3. Package the product/subject to suit the customer/reader. In reporting, instructing, or persuading, effective technical writers perform decision-making functions, and design/strategy functions. First, they decide what their information demonstrates or implies and why their readers want this information or how they will use it. Then, without distorting the facts, they design one or more presentations of the information, each of which is particularly suited to its intended purpose.

Technical writing, regardless of whether it involves a product, or is directed specifically to a customer or client, is always selling. To see the parallels between technical writing and selling, consider the characteristics of good salesmanship. An effective salesperson must know

- the product
- the potential customers
- how to convince customers that they need and should purchase the product.

Similarly, to communicate technical information on-line or in a report, letter, manual, or memo, writers must know

- the information—the facts and their implications
- the intended readers both as individuals and as members of an interpretive community—how they will operate in the situation requiring writing
- how to package the information so that the audience will "buy" it—understand what is important about it for *them*.

THE PLANNING PROCESS AND THE READER

If you think of technical writing as selling, you will remember to make the potential reader both as an individual and as part of an interpretive community central to the planning process that precedes writing. Just like a salesperson, you will organize the information you have to convey, not from your perspective but from the perspective of the people who are to read it. You will aim to answer their questions. You will arrange your information so that it satisfies the requirements of the job they have to do.

How does the salesperson adapt the features of the product to the needs and interests of the potential user? By learning and emphasizing the features of special concern to the prospect.

Example 1: Sales Brochures

The Better Box Company of Denver, Colorado produces a high quality line of cardboard boxes and charges accordingly for them. The written material its sales staff had been using to sell its product stressed quality. Over the last two years, however, they had been losing too many sales to lower-cost competitors. Finally, they asked former customers for specific descriptions of the products they had switched to for significantly lower prices. By listening to their audience, they discovered two key facts:

- In an inflationary economy the potential buyer's primary concern is saving money.
- The main competition was from box manufacturers that eliminate features or substitute inferior features creating a cheap product of limited durability.

Using this information, the Better Box Company rewrote its sales material, stressing the qualities of its boxes that made them more economical, in the long run than boxes with lower sales prices. The sales material discussed the advantages of stapled corners, for example, as opposed to glued corners.

The Better Box Company still cannot compete with companies whose strategy is gross underpricing. But now, it is losing far fewer customers and gaining new ones because its product descriptions are geared to its audience's needs and priorities.

Example 2: Applications Stories

To convince a company to purchase equipment, sales brochures must describe not only what the equipment does, but also how what it does can improve the operations of the prospective buyer. The applications story accomplishes this task by demonstrating how a company in a business comparable to the prospect's has used the product successfully. To write an applications story, the technical writer or applications engineer visits a company that has installed the product and interviews its managers and hands-on technicians. He/she collects diagrams, photographs, and product specifications and notes how the use of this product fits into the company's goals, values, and professional community interaction. Then the applications story writer produces a bulletin which by explaining why and how another company is using the product, suggests that similar companies might profit from it.

Figure 2.1 is an excerpt from an applications story used by Gould Instruments Division, Inc., to promote its direct writing recorders. The central premise of the brochure is that all large concerns trying to meet demands for energy conservation can, like Union Electric, achieve and maintain the highest possible efficiency from power plant equipment by using Gould direct-writing recorders for their monitoring.

A description of the product from the user/reader's point of view indicating by the example of another user the product's major applications is followed by:

- a description of how the product works.
- a flow-chart.
- a closing reiteration of the product's "tangible benefits."

CREATING THE READER'S INCENTIVE

Such a detailed description of a product would be of only passing interest to its intended readers, were it not preceded by the establishment of an incentive. My company is engaged in the operation of high-power generators, too; maybe I should consider Gould's direct writing recorders.

The literature of both the box company and the instruments manufacturer demonstrates situations where the writer and the reader are in a clear and specific seller/buyer relationship. The incentive for the writer to convey information from the reader's point of view is the writer's desire to promote and sell a product.

The same seller/buyer relationship is implicit in all technical writing. The engineer, manager, accountant, or psychologist who is most successful at communicating is the one who conveys information to the reader by adopting the reader's perspective and understanding the context and community in which the reader functions.

Technical writing is a selling job whether you are

- reporting information
- giving instructions
- persuading, proposing, or recommending.

REPORTING AS SELLING

Almost everyone reading this book has written and will continue to write lab reports, term papers, and other compendiums of information gathered from hands-on research, the work of others, or both. Most of us think of reporting

Gould Recorders Maintain Effective Power Plant Operation

With a growing concern for energy conservation and high fuel cost, Union Electric's 1200 megawatt Rush Island power plant has directed its efforts toward maintaining the highest possible generating efficiency. With a yearly fuel demand that approaches 2.5 million tons of coal, any minor inefficiency becomes cost prohibitive. To keep the customer's average cost per kilowatt hour down, thermal efficiency must be optimized. This is accomplished by effectively operating and maintaining the steam-generating system with highly trained technical personnel and Gould direct writing recorders.

Rush Island engineers learned how to attain the highest possible operating efficiency and maintain this efficiency from a power plant simulator at Westinghouse's test facility. This is where the control system is assembled and tested before being shipped to the customers site. Instructions on control system operation are simplified by the clear, crisp visual display on Gould 2800 direct writing recorders. Here engineers study the continuous traces on an amplitude vs. time record that provides a visual representation of control loop responses to power demands. Gaining experience under controlled, simulated conditions eliminates the expense of damaged plant equipment and prepares personnel for future emergencies.

Union Electric's Rush Island Plant depends on Gould 2600 Recorders for maintaining optimum steam-generation efficiency.

600 megawatt steam turbine-generator.
Exciter (left) A.C. Generator (center) Steam Turbine (right)

Figure 2.1 An applications story.

Coordinated Control System

The steam-generating coordinated control system at Rush Island functions like an analog computer. It is comprised of function cards that process system parameters into control loops. These loops monitor feedwater, air, fuel, pressure, temperature, and load fuel demand. Interconnecting them into a network of co-ordinating circuits, develops an information center that quickly responds to generation commands.

To avoid shutdowns that can cost over $30,000.00 an hour, maintenance must be undertaken with instruments that will dynamically record control signals accurately as they occur and interact with each other. Gould's 2600 direct writing recorder is an overwhelming choice by engineers where accurate reproduction of analog measurements are necessary. Providing a fast rise time on a rectilinear trace makes analysis easier and more meaningful when checking marginal operation, calibration of control loops, or solving an intermittent problem.

Steve Short, "Tech" Engineer, monitors reheat function cards at the information terminal center with Gould 2600 Recorder.

Reheat Temperature must be maintained at 1000° F. This is achieved by burner tilt and injection of water spray.

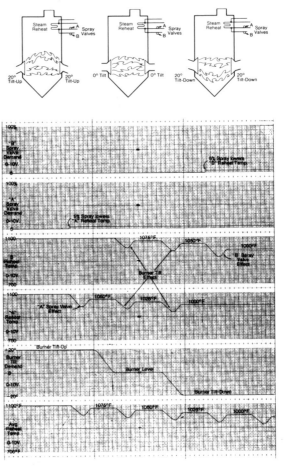

Figure 2.1 (*Continued*)

Typical Steam-Generation Power Plant

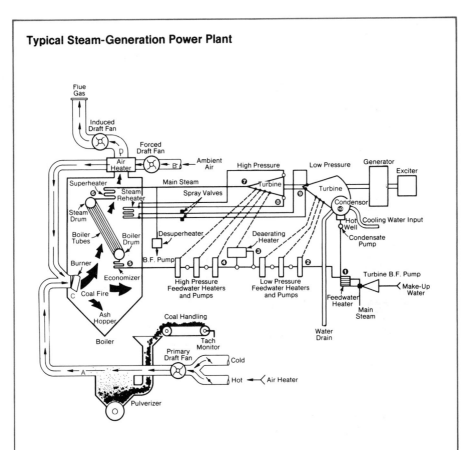

● **Air Flow Cycle**

(A) Powdered coal is drawn out of the pulverizer and blown into the burner by the primary draft fan. (B) Ambient air is heated and blown into the furnace. (C) Carbon in the coal combines with oxygen in the air creating a combustive reaction. (D) Induced draft fan creates slight negative boiler pressure and removes unwanted flue gas.

● **Feedwater Cycle**

(1) Make-up water replenishes any feedwater that is lost during the steam-generation process. (2) Extracted steam from the low pressure turbine heats the feedwater. (3) Oxygen is removed from the feedwater by deaeration. (4) Extracted steam from the high pressure turbine heats the feedwater. (5) Water enters economizer at 480° F. Feedwater will continue to increase in temperature as it passes through the economizer, boiler drum, boiler tubes and steam drum. (6) Steam enters the superheater and increases to 1005° F at 2400 PSI. (7) The main steam passes through the high pressure turbine producing torque to turn the generator and exciter. (8) Steam is then returned to the boiler and reheated to 1000° F at 200 PSI. (9) The reheated steam is passed through the low pressure turbine providing added torque for turning the generator and exciter. (10) The condensor turns the steam into water and the cycle starts again.

Plant Control

The air flow cycle and the feedwater cycle are controlled by a minimum of seven major control loops; all of which react with each other. A Gould multichannel direct writing recorder is the most effective method of visually displaying critical steam plant variables. This not only permits engineers to attain maximum thermal efficiency, but virtually eliminates the possibility of unscheduled boiler shutdowns.

Figure 2.1 *(Continued)*

Maintaining Effective Steam-Generation

Effective steam-generation is dependent on the control system for regulation of many variables, such as: primary draft fans, forced draft fans and induced draft fans. These must work effectively together to provide the required amount of air for complete coal combustion. Temperatures in the fan drafts must be high enough to prevent furnace cooling and assist in the removal of coal moisture. Furnace air pressure must be maintained within critical designed limits to avoid implosions or explosions. Feedwater temperatures and pressures are also adjusted to attain maximum thermal efficiency of the steam boiler and turbine combination.

These are only a few of the control variables that are constantly being regulated as generation-demand changes. Control room annunciators, meters, status lights and printers monitor these variables and indicate any deviation from their allowable limits. Whenever an indication of trouble appears, a technician is dispatched to the information terminal center. Two Gould 2600 direct writing recorders mounted on "equipment repair carts" provide read-out instrumentation that can monitor any twelve changing input-output parameters of function cards, or control loops, on continuous rectilinear traces. By analyzing these traces for proper

signal processing, response time or function sequencing, all malfunctions and marginal operations are quickly identified. After the control loop is repaired and calibrated, Gould's permanent chart record can be retained for future reference. This information becomes invaluable with time as control data accumulates; providing a ready reference for identification of virtually any component variation that effects optimum control system operation.

Conclusion

Conservation and cost savings are directly proportional to effective steam-generator control. To achieve this, every function card and control loop must operate within its designed limit. The complexity of accomplishing this task would be perplexing without the versatility of Gould direct writing recorders. Their accuracy and fast frequency response provides clear, concise data traces displayed on a permanent analog vs. time record. They are unmatched for ease of signal analysis and function card calibration. This tight control of system parameters is imperative since a 0.04% decrease in fuel efficiency means a $24,000.00 increase in annual fuel cost, or one hour of unscheduled down time can cost up to $30,000.00

Tangible Benefits

- **The Gould direct writing recorder provides an expedient method to help solve control system malfunctions and avoid unscheduled shutdowns that can cost up to $30,000.00 an hour.**
- **Each Gould recorder provides a direct cost savings that pays for itself annually for every 0.02% fuel inefficiency it prevents.**
- **The direct writing recorder provides the most accurate method for calibrating function cards and their dynamic response to each other, which enhances optimum control system efficiency.**
- **The Gould recorders provide an additional cost savings because of their ability to continuously monitor suspected intermittent problems, freeing personnel for other maintenance duties.**
- **Direct writing recorders provide clear, exact data traces on a permanent analog vs. time chart record, which is invaluable for training personnel on how a steam-generator reacts to various control commands.**

Gould Inc., Instruments Division
3631 Perkins Avenue, Cleveland, Ohio 44114
Telephone (216) 361-3315 TWX 810-421-8580
Bulletin 2019   Printed in U.S.A.

GOULD

An Electrical/Electronics Company

Figure 2.1 *(Continued)*

as stating the facts. We assume that the facts will "speak" to others just as they "spoke" to us as we accumulated them. If the report is handed in to an instructor and comes back with less than a "B," we are very surprised. We may be indignant because from our point of view we covered the subject in the way it should have been covered. But we can learn something about the nature of effective communication by analyzing the instructor's comments. For example,

> The paper contains some valuable insights on the subject, but does not fulfill the requirements of the assignment which was to write a comparison of two recent approaches to the problem.
>
> This paper contains an excellent analysis of your own approach to waste disposal, but fails to place it in the context of past research.
>
> Your argument is persuasive, but fails to anticipate the valid objections of the opposition.
>
> You retell too much and analyze too little.

Each of these comments indicates a communication situation in which the writing is at least partly ineffective because it has failed to meet the expectations of the reader. The reader cannot or will not "buy" the information in the way the writer has chosen to package it.

Wise students soon learn the interests, concerns, and expectations of their potential buyer—in this case, the instructor—and conform to those expectations in order to sell the next report as an "A" paper. This does not mean that students cannot express their own views, but rather that they must place them in a format that makes them accessible to the specific instructor for whom they are writing, and appropriate to an academic context. *In the community of business and industry, a comparable seller/buyer relationship exists between report writers and the customers, managers, peers, or subordinates with whom they communicate.*

Example

As a report writer you have information—how to improve a client's cash flow, how to automate an inspection system, what is causing the malfunction of the new basic oxygen furnace, and you have a reader or readers inside or outside your organization who have requested this information or who you believe will benefit from it. How do you communicate with them? Know

- who your reader or readers are
- why they requested the report
- how they will use the information.

Distance yourself from your work, to see, not why *you* are concerned with the number of defective lamp bases your company's producing, or the money wasted by not providing paid training in the use of newly installed equipment, but why it is of interest to the person/group who will read your report.

If you have any doubts about whether report writing is essentially selling and

For all types of persuasive technical writing, the basic seller/buyer dynamic is the same. You examine and analyze a body of information and come to certain conclusions regarding that information. You want to sell your readers on the same view of the information that you have. Once again, effective packaging depends on how well you anticipate your readers' interests, questions, and motives.

The following two samples were written in technical writing classes. In both cases, the object of the assignment was to re-think the components of a body of information in order to present it convincingly to two different groups of readers, each of whom would use the information for different purposes.

The first sample was written by a junior at Case Western Reserve University who has a double major in Management and Psychology. The second paper was written by Barbara Young, a metallurgical engineer at General Electric Lighting Business Group, Nela Park, Cleveland, Ohio.

Example 1: Explanation of Executive Assessment Program to Two Audiences

The subject of the undergraduate student's paper is the implementation of a psychological testing/assessment program for middle and upper-middle management. The company has decided to upgrade its hiring and promotion procedures by adding psychological testing to its assessment practices. A memo is written by a representative of the senior vice president of Personnel and Operations to the Management Committee appointed to make all decisions regarding implementation of the new testing program. This memo (Figure 2.2) aims to persuade the committee that (a) use of an outside testing firm as opposed to an in-house operation is more cost efficient, and (b) of the outside firms investigated, PRADCO offers the services that best suit the needs of this company.

The information on implementing psychological testing programs for executives has been presented from the point of view of the target audience—a decision-making management group—and the special concerns of the Management Committee have been taken into account. The writer has anticipated that in selecting an outside group to do this testing, the Management Committee will be most concerned with three issues:

- cost efficiency
- the relation of their division's activity to corporate operations
- assistance in making the sensitive area of psychological testing as acceptable as possible to the employees directly affected by it.

The major omission of this memo is the vague reference to "respected firms in the profession" other than PRADCO which were also considered. The writer should certainly name these firms and indicate why each was dropped from consideration. If readers are not told what the other choices are and why they are being eliminated, they will question the accuracy of the writer's judgment and wonder what other choices might be available. One or more of the committee members might have a specific company in mind and wonder whether or not the writer considered it and why.

If you have chosen between a number of options, tell your reader what the options were and why the others were eliminated before you go into detailed arguments defending what you did choose. If throughout the description of PRADCO, for example, your readers are wondering how it differs from BROWN, INC., their attention is diverted from the main points. Because they are distracted and confused by an unanswered question, they may not grasp the full significance of the information regardless of how clearly you have presented it.

The second document on the Psychological Testing program (Figure 2.3) is an open letter directed by the personnel office to middle and upper-middle management executives who will undergo the tests as part of their evaluation for promotion.

Once again, the writer has considered what motivates the readers' interest. The persons being asked to take these assessment tests will have the following questions:

• What do I have to do, and how much time does it take?
• How will the results of the tests be used in decisions regarding my retention, promotion, transfer, etc.?
• To whom will the results of my tests be available, and under what circumstances?

The letter from the personnel office attempts to address these sensitive questions. It shows executives that the tests can help their careers by identifying paths appropriate to their strengths and enabling them to cooperate with employers in identifying and overcoming weaknesses (paragraphs 5 and 6). It makes clear exactly what time and activities the tests involve (paragraph 4) and, also, where the results will be stored, who may examine them, and how long they will be valid (paragraph 7). Finally, it encourages the employees to ask questions.

Example 2: Two Versions of a New Equipment Proposal— to Decision-Makers; to Users

The subject of the second sample is a proposal for a new metallographic microscope for a Metals Laboratory. *Version 1* is directed at managers with minimal technical expertise who will make the final decision on purchase. The writer anticipates that their interest in a new microscope is primarily cost and benefits. She points out that by purchasing a "table-top" model of the proposed microscope, management can save $50,000 over the cost of the top-of-the-line version and still overcome most of the inadequacies of their present system. The comparison of different microscopes available, the recommendation, and the reasons are all presented with the concerns of management clearly in mind. (Spe-

To: The Management Committee

From: Special Assistant to Vice President of Personnel

Our company's Goals and Objectives for 1986 included implementation of a psychological testing and assessment program for key management-group-level executives. Responsibility for researching the various alternatives and presenting a proposal to the Management Committee by October 18th was assigned to the Senior Vice President, Personnel & Operations.

A thorough investigation revealed the two most common methods of assessment in industry to be (1) in-house, company-operated testing centers, and (2) use of a testing and assessment firm. An in-depth cost/benefit analysis prepared by our Personnel staff (Attachment I) indicates that the benefits derived from a company-operated assessment center would not justify the substantial costs involved in operating the center. Although the initial volume of assessments will be high, it is expected that all routine assessments will have been conducted within a year. Thereafter, only lower-level executives being considered for promotion and outside candidates being considered for key management positions will be assessed. Thus, the volume the first year may be as high as 250 assessments, but will drop after the first year to an estimated 30 to 50 assessments annually.

After extensive contact with respected firms in the profession, we are prepared to recommend contracting the services of the Personnel Research and Development Corporation (PRADCO), a Cleveland-based testing and assessment firm.

PRADCO offers several features which make it more attractive than the other firms we considered:

- PRADCO currently has the accounts of two large regional retail department store chains which provides them a base knowledge of our needs.
- PRADCO offers a personalized, informative process for the assessees through:
  - a detailed explanatory brochure given to each executive prior to the assessment;
  - assessment and feedback sessions for each executive conducted by Drs. Terry Owen and Stanley Rubin, both senior Principals of PRADCO.
- PRADCO is willing to work in collaboration with the Minneapolis-based firm of Hobert & Marting, a testing and assessment firm used by several other operating companies within our corporation. This commitment allows us to bring our program in line with the corporate inter-company synergy goal.
- PRADCO is conveniently located in Cleveland, only 15 minutes from Burke Lakefront Airport. In most cases, the assessment will be a one-day trip. Feedback sessions will be held by Drs. Owen and Rubin in our offices on a regular basis.

- A progressive fee schedule has been established by PRADCO as follows:
  - *Executive Assessment*—for middle management executives and outside candidates: $ 650
  - *Senior Executive Assessment*—for upper management and candidates for upper management positions (a more in-depth assessment requiring one full day at the PRADCO offices): $1000
  - Feedback session (per executive): $ 200

    Per diem (plus expenses) for roundtable sessions to interpret completed assessments: $2400

Considerations in Implementation

Psychological testing and assessment can be a highly sensitive issue. If not handled properly, it could defeat the purpose for which it is intended— development of the strongest management team possible. In response to this sensitivity, several points should be considered:

- Implementation of the program must be announced positively—perhaps, as an addition to the executive benefit package.
- Personal use of the assessment in career and long-range goal planning should be encouraged.
- Management's purpose for the assessment program should be explained: to match the right person with the right job in order to develop the strongest, most effective management team in retailing.
- Confidentiality should be assured. The written assessment will be reviewed exclusively by the CEO, the senior vice president of Personnal & Operations, and the senior officer of the division in which the individual works. The assessment will be kept in a locked file in the CEO's office for three to five years. It will *not* be put in the individual's personnel file.
- The individual will not be allowed to review his own assessment, but may ask questions that will be answered during the feedback session or anytime thereafter by the Senior Vice President of Personnel & Operations.
- The assessment will be used as a supplement to the personnel decision making process—it will not be *the* deciding factor.
- The fears and mystique associated with psychological testing and assessment should be alleviated by answering all questions openly and honestly and not turning this into a secretive, "whispered-about" process.

The success of this program will depend on our ability to communicate all of the above information to our management team effectively. Drs. Owen and Rubin will make a brief presentation at 9 a.m. tomorrow during our regularly scheduled Management Committee meeting. Afterwards, we will discuss how to proceed with implementation of this process before the end of the year.

Figure 2.2 Explanation of executive assessment to management.

DEAR _____ :

1   You may recall that a major challenge in our Company Goals and Objectives
for 1986 is a commitment to develop the most effective management team in
retailing. One component of this process is a testing and assessment program
designed to assist our executives in their personal and professional growth.

2   In keeping with this goal, we are pleased that we have retained the
services of the Personnel Research and Development Corporation (PRADCO)
of Cleveland, Ohio.

3   Drs. Terry Owen and Stanley Rubin, senior principals of PRADCO, have
been engaged to play a critical role in our management development efforts
through testing and assessment of our key management group. The
effectiveness of this program depends on the participation of each of you.

4   The time requirement on your part will be minimal, the benefits immense.
Sometime within the next six months, you will be asked to reserve a day to
go to Cleveland to participate in the assessment process. Prior to your trip,
you will receive a background data questionnaire and some reading material
which should be completed before your arrival at the PRADCO offices. Two to
four weeks after the assessment, Terry Owen or Stanley Rubin will visit your
office to provide you with feedback on your session in Cleveland and to
answer any questions you may have. Naturally, however, any questions you
have at any time may be directed to Don Brown, Senior Vice President,
Personnel & Operations, who will be happy to discuss the process with you
in detail.

5   The benefits you will receive from this process are innumerable. The
assessment is designed to identify not only your strengths, but areas in
which you may need further development. Your future annual Individual
Development Plans can be structured to meet your personal needs, rather
than following the generalized criteria currently established by the
corporation.

6   In addition, this process will assist us in matching the personal
characteristics of our managers with specific management positions as they
become available. Although an executive profile is not all-inclusive and
cannot be relied on as the sole decision-making factor, it certainly can be a
useful supplement to the decision-making process.

7   The written profile provided by PRADCO will be treated with the utmost
confidentiality and reviewed only by the chief executive Officer, the senior
vice president of Personnel & Operations, and your Pyramid Head. It will be
filed in a locked cabinet in the chief executive's office and will be retained for
three years, at which time it will be considered obsolete.

8   If you have any questions regarding this process now or in the future, feel
free to contact Don Brown or Peter Jones in the Personnel Office.

Figure 2.3   Explanation of executive assessment to participants.

1.

December 3, 1986

To: Joseph Burns
     Manager, Lab Operations

From: Barbara Young
       Senior Metallurgical Engineer

Re: Proposal for New Metallograph

The present needs of the Metals Laboratory have expanded beyond the capabilities of the Reichert metallograph now in use. Communications with the Reichert service representatives indicate that updating the system would be costly and without guarantee of satisfaction. As a result, several metallographic microscopes have been evaluated for replacement of the present microscope.

Two top-of-the-line metallographs, the Leitz MM6 and the Zeiss Axiomet are both excellent optically with wide versatility and high-quality photographic systems. The cost of each would be in the neighborhood of $80K. As an alternative, both Leitz and Zeiss have table-top metallographs which would work satisfactorily and would still have the high-quality lenses available on the more expensive models. The difference between the top-of-the-line and the table-top models is that the more expensive top-of-the-line models have a more sophisticated microscope mounting and more built-in components. The "table-top" models run in the neighborhood of $30K. Of the two, the Leitz Epivert is recommended as the better microscope system because of its broader capabilities and better camera system.

There are many restrictions in the present system, which would be alleviated by the acquisition of a new metallograph. The Reichert metallograph has a single low level light source which practically prohibits the use of color film or polarized light photography. There is no adaptation for 35 mm film, a feature which would reduce film costs and result in higher quality photographs. The lenses are not state-of-the-art in field flatness or light correction, and a nonstandard tube length does not permit updating. The camera screen is not parfocal with the oculars, requiring refocusing for photography. The camera shutter is faulty and often sticks. Objective lenses are singular and require stage repositioning every time one is changed.

A new metallograph would result not only in higher quality photographs and broader photographing capabilities, but also in a savings in film costs and operator time as well. A table-top model such as the Leitz Epivert would be a satisfactory alternative to a higher-priced top-of-the-line metallograph.

Figure 2.4   Proposal for new metallograph to decision-makers.

2.

December 3, 1986

To: All Metallurgical Engineers Who Use the Reichert Metallograph

From: Barbara Young
        Senior Metallographic Engineer

Re: Proposal for New Metallograph

    Several models of metallographic microscopes have been evaluated for acquisition, using demonstrations and available literature. As a result, I am recommending the Leitz Epivert microscope and the Vario-Orthomat camera system to replace the present metallograph in the Metals Laboratory. Criteria used in the selection were excellence in lens quality, illumination systems and camera, versatility, and price. Also important are the future add-on capabilities of the system so that as new techniques develop, the microscope can remain state-of-the-art. Selection of microscope accessories will be based on requirements of projects presently supported by the section and budget limitations.

    Of the microscopes considered, two models, the Leitz Epivert and the Zeiss ICM 405 table-top metallographs, stood out as having very high quality components. The ICM 405 has a unique compact design with a greater number of built-in components than the Epivert. The Zeiss model has a few drawbacks for which it was down-rated. Since the ICM 405 is still in the developmental stage, some accessories are not yet available. The camera system has a poorly designed film numbering system which works only by altering Polaroid film developers. Communications with other Ziess microscope owners indicate that service on equipment can be slow and less than satisfactory. The overall price is comparable to the Leitz model.

    The Leitz Epivert is excellent in quality of lenses and illumination systems, and has a broad assortment of components available for this model. Recent standardization of the Leitz microscope systems would ensure future updating capabilities. The most attractive part of the equipment is the Vario-Orthomat camera system which is far superior to the other systems evaluated. The system is fully automated with a zoom lens, a reticle projected by the camera, and a spot exposure meter for goof-proof photography. The Leitz Epivert would satisfactorily meet the needs of the laboratory.

Figure 2.5 Proposal for new metallograph to users.

cific costs and benefits that would normally be included in such a report have been omitted to protect company privacy.)

Version 2, addressed to co-workers of the metallurgical engineer who will use the equipment, demonstrates how very differently the same subject can be handled when the writer has the needs of a different audience in view. What questions will this hands-on audience raise as they read the description of the proposed new Leitz Epivert Microscope? Naturally, they will be concerned with the parts and capacities of the equipment and with its effects on each worker's projects and tasks. Therefore, in this version the writer is more precise about the equipment and its add-on capacities. Her second version would be even better adapted to the hands-on audience, if it mentioned projects currently being worked on with the old equipment and how the new equipment would make them easier or extend their potential.

SAME INFORMATION, DIFFERENT PACKAGING

Adapting the same body of information to the needs and uses of very different audiences is an everyday task in industry. The budget director has one reason for needing to know about the new electric coil you are developing; the foreman who will have to produce it on his existing assembly line has another. A cemetery monument dealer wants to know about your automatic sandblast equipment in terms of efficiency and profitability; the representative of Monument Workers Local 237 wants to know whether the equipment is designed with adequate safety features.

To understand this selection process, try visualing a body of information on a particular subject as a group of triangles. Suppose, for example, you are writing a feasibility report on automating the manual inspection process that your company uses to control product uniformity. Your available information falls into eight categories, each represented by one triangle (Figure 2.6). In writing

Figure 2.6   Eight separate information triangles.

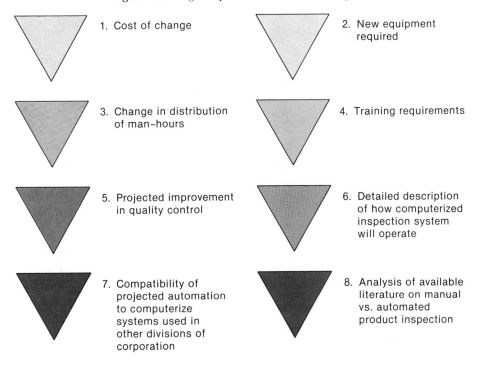

1. Cost of change

2. New equipment required

3. Change in distribution of man-hours

4. Training requirements

5. Projected improvement in quality control

6. Detailed description of how computerized inspection system will operate

7. Compatibility of projected automation to computerize systems used in other divisions of corporation

8. Analysis of available literature on manual vs. automated product inspection

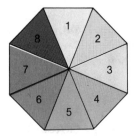

Figure 2.7   One way to structure the information in Figure 2.6.

your report, you will produce a document which arranges these eight triangles in a workable relationship (Figure 2.7).

As you can see, there are many ways in which the eight triangles (groups of information) can be related to each other to form a coherent whole (a clearly structured report). The shape or format you select, which items you discuss first, second, or last; which get detailed emphasis; and which, in some cases, will be omitted all are determined by the decisions you make regarding who will read the report and what it will be used to accomplish.

For the Director of Personnel and Training, for example, triangles 3 and 4 are vital; 1, 2, 5, 7 and 8 are of general interest and item 6 is possibly not necessary in any detail. A report directed at this reader would stress the human factor, preparation and handling of present and prospective employees, and the effect of automation on staff relations. Using the triangle analogy, a report prepared for this audience might appear in our triangle like the structure in Figure 2.8a. Other possible configurations are shown in Figures 2.8b, c, and d.

The packaging and re-packaging of the same body of information to make it suitable for use by different persons for different purposes is *not* distortion. The facts (triangles) remain the same. Certain components play a more or less prominent role in the design of the report—and where appropriate some components may disappear altogether. Design specifications which are vital to the persons who must build the equipment, for instance, are of little interest to the company's tax consultants who must determine the best way to appreciate the cost. Just as engineers frequently re-adjust the original design of a piece of equipment to accommodate the requirements of different users, so writers frequently redesign the presentation of their facts in consideration of the different purposes for which they will be read. By designing the information package to suit the reader, you do the best job possible of "selling" what you have to communicate.

EXERCISES

The following exercises are designed to give you practice in

    a. recognizing the perspectives of different audiences
    b. adapting the same body of information to the needs and expectations of different readers.

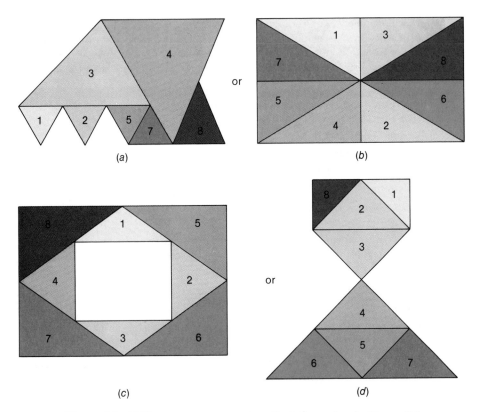

Figure 2.8 Different ways to structure the information in Figure 2.6.

1. Understanding how different people can see the same things differently is the key to successful communication. To help raise your consciousness, try to recall a traffic accident in which you were involved or which you witnessed. If you are fortunate enough never to have been in or seen a traffic accident, use one that you read about or saw on television, or make one up. Write *two* descriptions of the accident (include diagrams, if you wish), each one written by a different person for a different purpose. For example:

   a. from the person causing the accident to his insurance agent

   b. from a concerned citizen to the local newspaper using the incident to demonstrate a need for more stop signs, safety guards, etc.

   c. from a policemen filing the official accident report with his supervisor, and so on.

   Exchange papers with your fellow students and discuss what changes in facts, perspective, language, and attitude occur in the different versions, and why. If there are discrepancies in the facts, for instance, are they the result of wishful thinking, or misunderstanding?

2. To adapt the same information to different readers, you need to analyze the readers' relationships to the information and produce reports which satisfy each reader's needs. To practice this technique, do one or more of the following writing assignments. Then exchange papers with your fellow students in a workshop format. Pre-

tend to be the audiences to whom the reports are addressed, and decide whether each report presents the information in the way best suited to the target audience. Analyze what works, what does not work, and why. Individually, or in groups, rewrite any sections of your reports that do not meet the criteria of appropriateness for the intended reader:

a. A report on the advantages of supermarket banking

version 1: from a bank representative to a representative of a supermarket chain proposing the project

version 2: from a bank or supermarket representative to the ad agency that will design promotional material to "sell" the consumer on supermarket banking.

b. A description of a new product, such as an energy-saving lightbulb

version 1: in the form of a proposal to the company you want to adopt or produce your suggestion

version 2: in the form of an information sheet to regional sales managers who will have to introduce this new product to their sales staff and customers.

c. A description of a piece of equipment

version 1: in the form of a manual entry for mechanics who use and maintain the equipment

version 2: in the form of a recommendation to a non-technical manager that the company purchase this equipment to improve efficiency, solve labor problems, etc.

d. A proposal for a change in the curriculum requirements, or the syllabus of a course, at your school, or university, or a proposal for a change of procedures at your place of business:

version 1: in the form of a proposal to the person(s) or governing body that determines whether or not to make the change

version 2: in the form of a description of the change directed to those who will be immediately affected by it.

Patterns of Logic

In the technical writing process, writers often reverse the logical patterns they follow as researchers, problem-solvers and hands-on processors of information. They convert their patterns of logic—from specific facts to general conclusions—into the readers' pattern of logic—from general conclusions to specific facts. The technical writing process is not a *re-creation* of the logic the writer went through to accumulate information, but rather the *creation* of a logical pattern that the reader can follow to understand, accept, and utilize another person's work.

A difference in *patterns of logic* distinguishes *the problem-solving process* by which the technical writer acquires information to be written about first-hand, and *the reading process* by which the reader assesses this information. When scientists, engineers, or business analysts problem-solve, they move from specific facts to larger conceptions sustained by putting those facts together. They reach their conclusions last. But when readers attempt to understand the results of this work, they need to know the context or conclusions first, so that they can evaluate each fact based on what they have already been told it will demonstrate.

In this sense, the technical writing process is a process of reversal. To accommodate readers' patterns of logic, technical writers begin with their conclusions and work their way backward via analysis to the individual facts. They move from specifics to generalizations to achieve their results; they move from generalizations to specifics, from the top down, to report their results to others.

For example, a manager figuring out the best strategy for corporate planning gathers and analyzes the facts and draws a conclusion. As illustrated in Figure 3.1, the manager moves to the conclusion that his company's strategy over the next three years should be to maintain the produce business until the systems business has become self-sustaining. The boxes at the base of the pyramid rep-

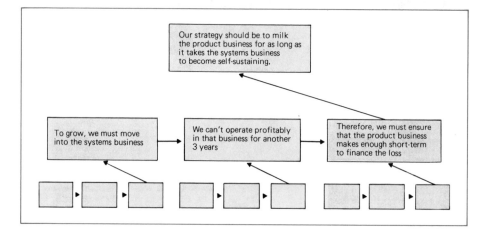

Figure 3.1 The researcher/writer's logic. (From Barbara Minto, "Deduction," *The Pyramid Principle* (London: Minto, 1975. Reprinted by permission.)

resent the most specific facts and figures that led the manager to the statement printed above them.

When the manager issues a companywide report announcing the company's projected strategy for the next three years, however, he does not re-create the step-by-step process that he went through to determine the most cost-efficient strategy. Instead, in his role as writer, he inverts the pyramid. That is, he begins with the conclusion—the key message of his report, and then moves down the pyramid to explain his reasons to the intended readers (Figure 3.2). Although problem solving begins at the bottom of the logic pyramid and moves up to the conclusion, the written report of the results of problem solving begins at the top of the logic pyramid and moves down toward increasingly detailed explanations.

Figure 3.2 The reader's logic. (Adapted from Minto, 1975.)

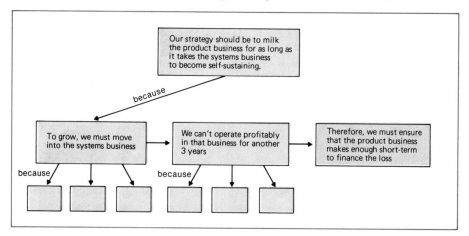

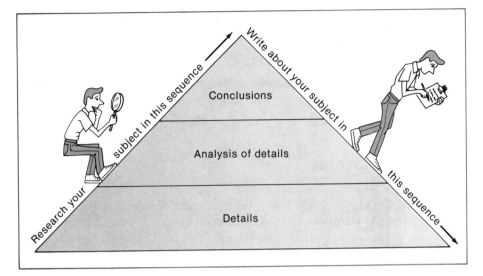

Figure 3.3 The inverted pyramid. (From Thomas P. Johnson, "Fast Functional Writing," *Chemical Engineering,* 76, June, 1969; reprinted in *Effective Communication for Engineers,* New York: McGraw-Hill, 1974.)

This reversal of logic, or movement up and down the pyramid which takes place when the problem solver becomes the technical communicator, is illustrated by Thomas P. Johnson (Figure 3.3) in an article in *Chemical Engineering* (reprinted in *Effective Communication for Engineers* [New York: McGraw-Hill, 1974, pp. 108–114]). As Johnson explains, it is natural for most problem solvers to assume that making sure their readers have a response to their information that is similar to their own response means making their report a faithful replay of this experience.

In fact, however, when you switch from your problem solving/investigating function to your report-writing function, you need to create for your readers not the same experience that you have had with the material but rather a *parallel* experience that will lead them to the same end.

Rearranging the sequence of events in a scientific experiment or problem-solving project in order to report them clearly to the intended reader is not, as Martin S. Peterson observes in *Scientific Thinking and Scientific Writing* (New York: Reinhold Publishing Corp., 1961), a distortion or betrayal of scientific truth. Instead, it is the responsibility of the expert to transpose her/his thinking into a format intelligible to the proposed recipients.

THE VIRTUES OF AN OPENING CONCLUSION

Why, then, must you begin just about every technical communication that you write with an opening summary of your conclusions?

- First, because your readers cannot understand the significance of the specific details of your work unless you give them an opening context to which they can relate those details.

- Second, because if you do not provide a clear context for your facts, your readers will guess at, or impose their own context to make sense of what they are reading. Unfortunately, if you do not provide the necessary guidance, the sense the readers make out of your facts and figures may not be the sense you intended.

THE READER AS CREATOR

To test these claims about readers' requirements, think about how you behave when you are reading technical or specialized information. What do you need from the writer in order to make sense of what you are reading?

Suppose, for example, you are presented with the comparison chart in Table 3.1, a competitive analysis of two different models of flatbed recorders. Disregard for the moment how much you know about the operation of these recorders and each of the criteria listed. What is the first thing you do as a reader when confronted with an unexplained group of facts?

Chances are that you will try to impose a structure and a context. Once you have established in your mind that this is a comparison table in which the performance of two products, ABC 123 and the XYZ 456, is being compared based on each of the criteria listed in the left-hand column, you will attempt to interpret what each of the comparisons demonstrates. After going through the entire table, you will probably try to draw some conclusions about the information you have accumulated—either that one product clearly outperforms the other; or that they are essentially the same; or that each has different strengths and weaknesses.

In almost any technical reading situation, the readers, given a body of information without a writer-imposed context will, of necessity, create their own. If the writer has not preceded the facts and figures with a firm suggestion of what they should demonstrate, readers will not be certain of the reason that each of the isolated details are significant to them. Readers cannot process information without relating it to a purpose, message, or situation. A moment ago, for example, you did not look at the chart in Table 3.1 just for the facts themselves; you looked at it to see what reading it would tell you about how readers process information when no context has been provided. You knew in what context to process the reading of this comparison, because the writer of this book provided the context beforehand.

Table 3.1 is preceded by a memo from a representative of ABC's Marketing Department to the materials sales manager. Its purpose is to explain the significance of the information in the table to the sales representatives, who must be able to discuss knowledgeably with potential purchasers and present accounts the difference between ABC's products and those offered by their competitors. The sales staff is often questioned in detail by customers about new products or

Table 3.1   Competitive Analysis
ABC Model 123 versus XYZ Model 456

	ABC 123	XYZ 456
Response Time	350 ms	500 ms
Inaccuracy	±.2% f.s. (Including nonlinearity & deadband)	±.2% f.s. on 500 mV range at 25°C
Deadband		±.1% f.s.
Remote Pen Lift	Standard	Standard
Chart Drive	Crystal Oscillator	Crystal Oscillator
Variable Span Control	Standard	Standard
Z-fold Paper	Yes	Yes
Chart Speeds	2, 4, 10, 20, 40 cm/min & cm/hr	2, 6, 20, 60 cm/min & cm/hr
Pen Position Control	Yes	Yes
Input Ranges	(13 ranges) 1,2,5,10,50,100,200,500 mV; 1,2,5,10 V	(12 ranges) 10,20,50,100,200,500 mV; 1,2,5,10,20,50 V
Input Impedance	1MEG ohm on all ranges	1MEG ohm on all ranges
Common Mode Rejection	120 db	120 db
Power Requirements	115/230 V—50–400 Hz	90 to 110 V, 100 to 130 V, 180 to 220 V, or 200 to 250 VAC (Must be specified) 48 to 63 Hz
Remote Chart Drive	Standard	Option (price not available yet)
Proportional Chart Drive	Standard	Option (price not available yet)
Event Marker	Left Hand Standard	Option (price not available yet)
Remote Pen Lift	Standard	Option (price not available yet)
Weight	1-pen 7.5 Kg (16.5 lb) 2-pen 8.9 Kg (19.5 lb)	1-pen 8Kg (17.6 lb) 2-pen 9 Kg (19.8 lb)
Dimensions	16.75″ W × 5.87″ H × 14.75″ D (42.6 cmW × 15 cmH × 37.5 cmD)	17.125″ W × 13.875″ D × 7.652″ H (43.5 cmW × 35.3 cmD × 19.2 cmH)

new versions of products that appear to offer more features or the same features at a better price. A memo addressed to the manager of these sales representatives, which might ultimately be distributed to the representatives themselves, would need to highlight the main points revealed by the table and make them immediately clear to personnel who are looking at the product not for its technical attributes, but for how these features affect the product's saleability as compared with others in the same market.

Try reading the memo (Figure 3.4) from the point of view of an ABC sales representative. See if it gives you the information you need to understand the table and to explain the differences to your customers. If you read the memo

To: All Sales Personnel
From: Marian Turner
Date: 4-2-85
Subject: Competitive Analysis—XYZ Model 456 Versus ABC 123.

The XYZ Model 456 flatbed recorder is a newly released low-cost strip chart marketed towards OEM and end users. It features a non-contact ultrasonic pen position transducer and brushless DC servo motor. The servo mechanism of both models is sealed for maintenance free, long-term reliability. However, our competitor cannot meet the performance criteria nor the standard features available with the ABC 123. .

Priced at $712. and $1220. for a 1- and 2-pen, respectively, the 456 appears to be the better buy; but when you consider features and performance, it becomes obvious that it is not.

Preliminary observations showed that the ABC 123 has no apparent weaknesses compared with XYZ 456 and that XYZ 456 has no strengths over our ABC 123.

ABC's selling features versus XYZ 456:

- Faster response time
- Better accuracy
- More standard features
- More chart speeds
- More useable input ranges (XYZ 456 lowest range is 10 mV)

A comparison of specs is attached.

Remember, ABC has the *best* strip chart on the market!

Attachment

Figure 3.4   Memo to sales representatives. (Adapted from instruments company's sales division memos.)

from the perspective of the sales representative, the moment you realize that it concerns a recently released competitor product, what is the first question that runs through your mind? Probably, what feature, if any, does it offer that might draw present and potential customers away from the ABC 123? Your second question will naturally follow: if it does have any preferable capacities or advantages, can I offset these by any significant disadvantages?

In other words, the recipients of this memo will be reading it solely to determine whether and how the introduction into the flatbed recorder market of the XYZ 456 affects their sales strategy. If the memo outlines and analyzes the infor-

The XYZ 456 flatbed recorder is a newly released low-cost strip chart motivated towards OEM and end users. A comparison of specs (see attached) between the ABC 123 and the XYZ 456 demonstrates that the ABC 123 excels the XYZ 456 in five categories and does not have any significant deficiencies in comparison with the XYZ 456.

Although the XYZ 456 features a non-contact ultrasonic pen position transducer and brushless DC servomotor not available on the ABC 123, both servo mechanisms are sealed for maintenance-free, long-term reliability.

Most important, the XYZ 456 cannot meet the performance criteria or the standard features available with the ABC 123. As the attached table shows, the ABC 123 compared to the XYZ 456 has:

- faster response time
- better accuracy
- more standard features
- more chart speeds
- more useable input ranges.

As you will be able to demonstrate to your clients, the ABC 123 is still the best instrument of its kind in this price range.

Figure 3.5   Revised memo.

mation contained in more detail in the table in such a way as to anticipate and answer questions that this subject will raise in the minds of its intended readers, then the memo is an example of successful technical communication.

What do you think?

Initially, participants in an in-house technical writing seminar rated this memo as adequate. However, as they talked about ways to improve it, they realized that the main point was difficult to understand and there appeared to be contradictions.

In the second paragraph, it is not clear whether the 456 features described in the first sentence are unique to the XYZ product or not. If the ABC 123 does not offer these features, then how can the statement in paragraph 3, that the XYZ 456 has no strengths over the ABC 123 be true? Should this statement be modified to "significant strengths"?

For these and other reasons, the participants in the writing seminar decided that the memo needed an opening context from which the rest of the information would follow. In a group workshop, they rewrote the memo (Figure 3.5).

Notice that in this second version very little of the information included has been changed. Instead, the relationship of the pieces of information to each other and to the concerns of the target reader have been clarified by the subordination of pertinent facts to an opening statement of context. This version of the memo is an improvement in the packaging of the information because it addresses the needs of the reader more directly.

As a reader, you can see clearly that it is much easier to grasp the significance of related items of information if the writer precedes them with an opening conclusion that summarizes a) on what basis the items are related or grouped, and b) the main reasons the relationship is significant from the reader's perspective.

As a writer you need to provide your reader with the same opening context that all readers require to process information presented to them by others.

WHY WRITERS SOMETIMES OMIT A CONTEXT

In your role as a technical writer, it is easy to avoid putting the conclusion first— Why? Sometimes, the writer does not give away the conclusion first, because the writer does not know it. Be honest with yourself. If you are not divulging the ending because you haven't digested and analyzed the material enough to know what it demonstrates, you are not really ready to write. If, pressed by time, you dash off a report that reads like the index of an encyclopedia, you are helping no one. The reader, who is usually your superior, will not get the information needed and will consider you incompetent, regardless of how hard you've worked or how much you know about your subject.

Sometimes, technical writers see themselves as the authors of mystery novels. They fear that if they tell "whodunnit" upfront, no one will read any further— or those who do read further will be bored. It's fun to be kept in suspense in your leisure reading. But if you were in charge of a bottling plant with a faulty bottle washer that was rejecting or breaking every fourth empty bottle, would you enjoy reading through several pages of testing procedure descriptions in a quality control report before you learned what the quality control engineer thought was wrong with the washing unit and how much time and money would be needed to fix it?

Writing in industry means converting the information you have acquired while doing your job into a message that is quickly accessible to your audience to help them perform their jobs. Whether your message consists of 3 lines or 300 pages, it should begin with a conclusion.

THREE LOGICAL ANALYSES

The remainder of this chapter examines three writing samples, one by an undergraduate accounting major, one by a steel company engineer and one by an electrical engineer. Our primary concerns will be:

- to determine the logical structure of each report—from specific to general, or from general to specific
- to analyze how each of these logical structures affects the usefulness and accessibility of the information to the intended readers

- to discuss what, if any, changes might be made in the logical structures to improve or increase communication effectiveness.

Writing Sample 1

Figure 3.6 is a memo written by an accounting major in her junior year who works part-time for an accounting firm. How is this memo structured? Is the reader given an opening context which clarifies how individual points of information are related, or does the reader perceive the meaning bit by bit? Is the communication successful?

The first reactions of the writer's classmates were objections to the inappropriately accusatory tone of the opening. They recommended illustrations to clarify the office layout and the proposed changes. But gradually, they noticed two more fundamental problems of logic:

- that the real priority message (from the reader's point of view), the conclusion, is not stated until the end, and not fully articulated even then,
- that the specific recommendations are not logically grouped so that the reader can grasp immediately their relationship to the priority message and to each other.

The class as a group rewrote the opening of the proposal (Figure 3.7).

By starting with this summary, the proposal prepares the reader to receive specific recommendations about where the file cabinets and other items should be more favorably and with a clear understanding of purpose.

Writing Sample 2

Figure 3.8 is written by a metallurgical engineer to inform other metallurgical engineers of the special design problems involved in tailoring steel piping for use in the drilling of deep oil wells. The author's peers all agreed that the memo did not provide its readers with a clear opening context. They felt that most of the section labeled "Introduction" was irrelevant to the intended audience, since engineers engaged in this project would already know the background of the problem. The last two sentences of the introduction do really introduce the report, but unfortunately, they mislead the reader regarding its subject. As it turns out, this report does not describe the development of down-well tubing, but rather three prevalent problems in the design of down-well tubing and what is being done about each of those problems.

Section II, labeled "Oil-Well Servicing Conditions and Environmental Factors," is actually a description of the three problems—H_2S Cracking, Low Cycle Fatigue, and Low Resistance to Softening.

Section III, labeled "Design Criterion Used in Developing Down-Well Tubing," is actually a discussion of what was or might be done to alleviate each of these problems. To add to the readers' confusion, the three problems are discussed in one order in Section II and in a different order in Section III.

April 13, 1987
To: The Officer Manager
From: Sheila Warren
Re: Arrangement of Office Furniture and Equipment

An accounting office should have a professional and business-like atmosphere. The rooms should be neither cluttered nor empty. Equipment and supplies must be within easy access for all employees.

Currently, this office is set up in an unbusiness-like, inefficient, and haphazard manner. The file cabinets are thrown in the middle of the staff room, where they merely serve to get in the way. They are not close enough to the staff members so that these people can remain seated while using them; the files are at the opposite end of the office from the secretary's office and the conference room. This condition makes it very hard to answer questions on the phone without leaving the client on hold for a long period of time. It can also cause problems during a conference, if a file is needed immediately.

I propose moving the files from the middle of the room to the wall separating the conference room and the staff room. In this location, the files will be out of the way, but they will also be within easy reach of everyone who uses them.

A major source of inefficiency in the staff room is the location of the desks. Of the three main desks, only one has sufficient lighting. Without enough light, staffers have to concentrate on making their figures dark enough instead of making sure they are correct. In order to correct this problem, the desks may be moved from the left side to the right side of the office. If this is done, each person will be working directly under an overhead light and next to a window.

The last cause of problems in the staff room is the Xerox machine. As it is currently situated, there is no place to collate the copies as they are made. This is annoying and a waste of time. If the Xerox is moved to the left wall, where the desks are currently placed, there will be room to place a small desk or table next to the machine for collation purposes. The Xerox will also be in closer proximity to the file cabinets, allowing files to be put away immediately.

The reception room should also be rearranged. As the secretary's desk is currently placed, she cannot look directly up from her work to greet clients as they enter the office unless she is typing. Since the majority of her work does not involve typing, her desk, not the typing extension, should be facing the doors. This can be achieved by putting the desk in the left-hand corner of the room closest to the street. The sofas and chairs can be moved to where the desk was, allowing the clients not only a friendly greeting from the secretary, but also a more pleasant atmosphere to wait in. As the sofas are currently situated, the waiting client can look only at magazines or a bookshelf. If this furniture was moved to the desk's current position, the clients would also have the option of studying the pictures or gazing out the window.

As it is currently set up, the conference room has a professional quality. The table and chairs are in comfortable positions, and the chairs are also set up in a manner that does not allow a client's mind to drift away from the conference.

If the above suggestions are carried out, I feel that the office will run much smoother and more efficiently. The clients will receive a more favorable impression of the office, and consequently a more favorable impression of the business itself.

Figure 3.6 Proposal to redesign an accounting office.

The real message of this report and the relation of individual facts to the central message can be made readily apparent to the reader by restructuring. The writer must provide an opening context and subordinate each component to the clearly defined priority message. For example,

There are three oil-well servicing conditions that engineers need to consider in designing down-well tubes for use in deep oil well drilling. They are: 1) Low Cycle Fatigue, 2) Resistance to Softening, and 3) H_2S Cracking. The development of an intercritically batch-annealed "dual phase" skelp appears to solve the first two of these problems. The new design increases fatigue resistance by increasing ductility, and decreases resistance to softening by replacing precipitation hardening with dispersion and solid solution hardening. The effect of this design change on H_2S Cracking is not yet known and will have to be tested in the field.

Directed at the non-technical manager, this one-paragraph progress report might be all the writer would need to inform the reader sufficiently. Directed at

Figure 3.7 Revised proposal.

There are two major criteria for determining the best arrangement of furniture and equipment in an accounting office: the arrangement should (1) enable the accountants and support staff to function at optimum efficiency, and (2) give an impression of efficiency and professionalism to clients and visitors.

At present, the conference room of our office fulfills both these criteria. The furniture is comfortable and is arranged to discourage distractions. But the staff and reception rooms need improving. To increase the efficiency and professional appearance of our entire office, I propose two changes:

- *in the staff room*—relocation of desks, files, and machines in order to allow minimum employee traffic and maximum use of adequate lighting facilities;
- *in the reception room*—rearrangement of desk, sofas, and chairs to enhance the atmosphere for waiting clients.

October 12, 1985

To: Metallurgical Engineers Involved in Developing Steels Tailored for Use in Oil
    Fields

From: John Sturbridge, Senior Metallurgical Engineer

Re: Down-Well Tubing

I. Introduction

In recent years an energy crisis has developed whose central focus is
petroleum products. The value of crude oil has risen steadily, and this has
spurred the development of new techniques and technologies in all areas
dealing with the production and utilization of these products, many of them
outside the oil companies themselves. One of the areas undergoing change is
the mining of crude oil. The increased demand for petroleum products has
created a need to tap oil reserves which before had been considered unprofitable
due to the great depths to which drilling would have to take place.
Additionally, the yield and efficiency in crude oil production from oil wells has
been improved by utilization of new oil well servicing techniques. This report
describes the development of a high-strength steel-coiled tubular product
which has resulted in improved oil well servicing, particularly for deeper oil
wells.

II. Oil Well Servicing Conditions and Environmental Factors
    A. H_2S Cracking

Oil Well servicing is done to facilitate efficient production of crude oil from a
well. However, the environment which the tubing must withstand can change
from well to well. This is of particular importance in the consideration of a
product's resistance to H_2S cracking. As oil wells become deeper, the probability
of having a sour (H_2S-bearing) versus a sweet (non-H_2S-bearing) well increases.
But in order to service deeper oil wells, higher strength service tubing must be
utilized. This only aggravates the problem, since generally a material's
resistance to H_2S cracking decreases as strength of hardness increases.

To guard against failures caused by H_2S cracking, the National Association of
Corrosion Engineers has established NACE specification MR-01-75. All tubing
used in sour service must meet each of two criteria:

    1. Hardness must not exceed Rc22.
    2. Nickel contents of the alloy shall not exceed 1%.

B. Low Cycle Fatigue

Due to the nature of the servicing conditions, the tubing must have excellent low cycle fatigue resistance. This is due to the reversed plastic strain conditions the tubing is subjected to with each cycle in and out of a well. A design criterion of at least 50 cycles to failure has been established.

C. Resistance to Softening

The cyclic nature of the service conditions can also yield some cyclic softening with subsequent reduction of yield strength. However, it is believed that the main purpose of the yield strength should be to maintain dimensional stability of exterior tubing dimensions and that failure in service will be governed by the tensile strength which is not subject to cyclic softening to the degree that yield strength is.

III. Design Criterion Used in Developing Down-Well Tubing

In the developing of down-well tubing, the following design considerations were considered:

1. Since low cycle fatigue resistance is related most often to the inherent ductility of a material, it was deemed desireable to optimize the strength-elongation relationship.

2. It was assumed that utilization of dispersion hardening and solid solution hardening mechanisms should be maximized at the expense of precipitation-hardening and cold work (via partial recrystallization of cold-rolled annealed skelp). This assumption is based on the fact that the latter two strengthening mechanisms are more subject to cyclic softening.

These considerations led to the development of an intercritically batch-annealed "dual phase" skelp since improved ductility was realized while employing solution and dispersion hardening mechanisms. The H_2S cracking resistance of this approach is unknown. There are microstructural considerations which are both favorable as well as unfavorable in their implications of predicting resistance to H_2S cracking. Only servicing in the field can readily answer this question.

If one wishes to look for a "silver lining" in the energy crisis cloud, it is interesting to note that development of batch-annealed, dual-phase steel was at least partially generated by the oil shortage. This development is now being used to help manufacture lighter, more fuel efficient automobiles.

Figure 3.8 Explanation of down-well tubing to metallurgical engineers.

February 21, 1984
To: EFG Electric
From: William Kramer, Electrical Engineer II
Subject: Kuro Company Basing Evaluation

Kuro has returned the dummy lamps and bases which were supplied earlier this year. Various soldering and welding operations were performed on the bases and the following data is an evaluation of these bases. The evaluation is by visual inspection only, but metallographic analysis should be available in about a week.

I. *Side Solder*—60 threaded seal lamps were examined having side solder. 34 used the (EYE) solder with composition 4SN, 96P_b and 26 used the (CRAMCO) solder with composition 5SN, 93 P_b, 2Ag. Results as follows:

    1—Solder on base threads   CRAMCO
    2—Slight cracking of Base Shell   CRAMCO
    3—FLUX deposit on jacket   EYE
    1—Questionable wetting   EYE

The slight cracking in the base shell may possibly be due to stress corrosion cracking. The flux deposit on the jacket was in-line with the arc tube and therefore would require cleaning. However, on almost all 60 lamps some degree of flux extended 15 mm above the base onto the jacket. This may or may not require a cleaning operation. In general, the lamps were acceptable with both the EYE and CRAMCO side solder.

II. *Soldered Eyelet*—14 lamps were examined using the CRAMCO solder (same as side solder). Results were:

    3—Considerable glass base cracking*
    2—Poor wetting of solder to eyelet
    4—Questionable wetting

In general, the appearance of the soldered eyelets seemed to be poor. In my opinion the bases did not even compare to the Canadian soldered bases which were evaluated earlier. However, since no specs were given to Kuro Company on soldering, it may be that the quality could be improved.

III. *Plasma-Welded Eyelet*—46 lamps were examined with plasma welded eyelets and four different welding conditions. Categorizing these conditions from best to worst with the major defect of each concluded:

    A—Insufficient data, 1 lamp tested
    B—5 of 11/glass base cracking*
    C—17 of 28/glass base cracks*, 10-voids in the weld, 8-burned thru eyelets.
    D—6 of 6 with voids in the weld at the eyelet hole.

The above results would indicate that plasma welding needs considerable refinement before usage.

IV. *Plasma Welding Using Copper Ring*—26 G.E. bases using an EYE-Lead Wire (Monel: 65% Cu + 35% N_l) and a copper ring showed improved results over our N_l Plated Fe Lead wire with plasma welding:

    8—Small voids in weld
    10—Glass base cracks*
    1—Burn thru on eyelet

A significant improvement in weld quality was achieved by using the copper ring along with the EYE Lead Wire. The small voids in the weld are probably not detrimental to the lamp.

(*These bases had 6 or more cracks in glass which extended from the eyelet to the shell. This may or may not be significant to lamp quality, however, it was noted.)

V. *Conclusions*
Acceptable side soldering of the thread seal lamp can be achieved with either the CRAMCO or EYE solder, at least by visual inspection. It may be possible to get an acceptable eyelet soldering operation developed with further testing. Plasma welding of the lead to the eyelet was totally unacceptable. However, plasma welding using the copper ring technique is very promising.

Figure 3.9   Basing evaluation memo. (Adapted from student paper in GE Manufacturing Studies writing course.)

a more technical reader (persons designing or building this tubing), this paragraph would provide the proper opening context for the more detailed discussion of each problem and how each is affected by the development of the new batch-annealed, dual-phase steel.

Writing Sample 3

Figure 3.9 reports on the problem of selecting the best of several ways to solder and weld lamps to bases. The writer, an electrical engineer, is describing the results of the visual inspection of products on which the Kuro Company at the request of a lamp-manufacturing client performed four different soldering and welding operations.

A group of engineers discussing this report in a workshop all agreed that the information was delivered in a cumulative, piece by piece style, and lacked an opening summary that would place each item of information in a clear context.

KURO COMPANY BASING EVALUATION (*revised version*)

Opening
Summary

    In May, 1984 a supply of dummy lamps and bases was sent to Kuro so that they could perform four types of soldering and welding operations on the bases for GE evaluation. The finished lamps have been returned, and based on evaluation by visual inspection only, *plasma welding using the copper ring technique* seems the most promising. Of the other three alternatives, *side solder* with either EYE or CRAMCO is acceptable despite minor imperfections; *eyelet soldering* is not presently satisfactory, but might be brought up to standard if specs are issued to Kuro; and *plasma welding of the lead to the eyelet* is totally unacceptable.

Scope

    This feasibility report provides a detailed analysis of the strengths and weaknesses of the four methods based on visual inspection. An evaluation based on metallographic analysis by G. I. Thomasson should be available in about a week.

Facts and Analysis

    I. *Side Solder*—60 lamps having side solder (34 EYE, 26 CRAMCO) were inspected. Despite some minor problems, all 64 were acceptable:

# of lamps examined	observations	comments
34 EYE solder w. composition 4SN, 96P_b	slight cracking in base shell, 2 lamps (CRAMCO)	may be due to stress corrosion cracking
26 CRAMCO w. composition 5SN, 93 P_b, 2Ag	solder on base threads, 1 lamp (EYE)	
	FLUX deposit on jacket, 3 lamps (EYE)	in line with arc tube and would require cleaning
	questionable wetting, 1 lamp (EYE)	

Note: On almost all 60 lamps some degree of flux extended 15 mm above the base onto the jacket. A cleaning operation may or may not be required. In general, the lamps were acceptable with both the EYE and CRAMCO side solder.

(Follow through on same pattern with other three methods and use same conclusion as in original.)

Figure 3.10   Revised memo.

For example, after reading the section on side soldering, the workshop group observed that they do not know yet how many other and what kind of soldering and welding operations they will be comparing it to further on in the report, and they do not know all of the criteria for evaluation. When they read that one CRAMCO base had solder on the base threads, they do not yet know if this is a significant defect in itself or only in comparison with the defects observed when other soldering methods are used. Until readers come to the last paragraph, they do not have all of the information they need to understand each of the facts preceding that last paragraph.

In fact, as you may know from your own experience, many readers when encountering information for which there is no explanatory context, will automatically turn to the last page or last paragraph and read the conclusion *first* on their own initiative in search of a context.

When the writer has reversed the logic for the reader and has opened with a conclusion under which all of the supporting information is logically subsumed, the reader's job of comprehending is made much easier. It is much more likely that readers will get from the information both what the writer wants them to get and what is most useful for them to have.

The group's revised version of this report (Figure 3.10) shows that *when the material is introduced properly by a summarizing umbrella, the report becomes considerably shorter.* The writers found that after making the key points in the first paragraph, they could present all the supporting information in table form with only one or two sentences of explanation under each type of soldering or welding. In other words, converting information from your investigation-phase logic into a logical order more conducive to reader comprehension simplifies and shortens both writing time and reading time.

EXERCISES

The following exercises are designed to give you practice in experiencing the difference between the logic of problem solving and the logic of writing, and in transposing material from one logical structure to the other.

1. Your guidance counselor or major advisor at school or your supervisor at work has asked you to prepare a description of your academic or career goals for the next two years.
   Step 1: Brainstorm. Off the top of your head, list all the things you want to do—courses you want to take, trips, books you want to read, people you want to make contact with, independent projects you want to complete, and so on. Also list, as they come into your head, your goals or reasons for wanting to do these things. As you go through this exercise, you will be accumulating and processing information.
   Step 2: Organize your material. When you have run out of thoughts, organize and write up your material. That is, produce an opening conclusion which summarizes your main goals and the major ways in which you hope to achieve them, and which leads your reader into the supporting details that will follow. Try to organize the subordinate details into categories to facilitate your readers' understanding.

2. *Step 1: Brainstorm.* You are doing a comparative analysis of three products, processes, or career opportunities about which you are well-informed. Create a list of criteria that you would use to make a choice or help your reader make a choice between the items being compared. For example, which of three home computers to buy, which of three graduate schools to attend, which of three exercise programs to follow. If you have the material from class or from your job for making a more technical comparison, of different chemical processes, for example, by all means do so.

Step 2: Evaluate the Choices. Having compiled a full list of criteria, evaluate each of the three choices next to each of the criteria in the comparison frame below:

Points of Comparison	Choice A	B	C

Step 3: Determine a Preference. Evaluate the information you have compiled to determine which if any of the three choices is clearly preferable, and why.

Step 4: Present Results. Write an essay in which you present the results of your comparison and the supporting evidence. Be certain that your essay begins with an opening conclusion that states which choice is preferable, summarizes the main reasons, and sets up a context for the detailed supporting evidence that will follow. Get feedback on your essays from several different readers by discussing them in class in small workshop groups.

The Process: Before, During, and After

CHAPTER 4

The Writing Process: Before You Write

The planning phase of the technical writing process consists of:

* listing everything you know about the audience and situation;
* listing or outlining everything you know about the subject;
* analyzing your reader-description list and information-description list to determine an opening conclusion and an order of presentation of supporting facts.

DECISION MAKING

The first decision you make (unless it is made for you) is that a situation exists which makes a technical communication necessary. It may be that part of your regular responsibility is a monthly or bimonthly progress report. It may be that someone inside or outside your division (usually your superior) has become curious about your research or your group's functions and has requested a descriptive report. Maybe you yourself have decided that (a) you have done enough research on a new product or process to propose it formally, or (b) upper management will ignore what you are doing unless you put it in writing, or (c) there is a personnel, equipment, or operational problem or some new development that others outside your immediate area need to know about in writing.

Your next step is to decide what your readers' perspective on your material will be. What are their areas of expertise? their job responsibilities; will they affect and/or be affected by the subject matter of your report? How does the content of your communication affect the performance of their jobs, the advancement of their careers?

J. C. Mathes and Dwight Stevenson in *Designing Technical Reports* (Indiana: The Bobbs-Merrill Co., Inc., 1976) suggest a method of audience analysis focusing on creation of an "egocentric organization chart" and the maintenance of individual files on the educational background, special interests, and so on of each person inside or outside your organization to whom you write regularly. They also discuss how to distinguish your decision-making or primary reader from others. It is the decision-making reader or group to whom you direct your writing.

If you write regularly for the same people or types of readers, keeping and updating elaborate charts and files of your audience, while time-consuming, definitely simplifies your writing task in the long run. Not only do you need to understand these individuals, you also need to know how they interact as a group and how they interact with you.

To start with, however, all you really need is a basic sense of other, an ability to imagine how someone in a particular position other than your own would most likely react to the subject of your writing.

THE READER'S PERSPECTIVE

If you are writing to a professor to protest your final grade in a course, which of these lines of argument do you think would be most effective?:

- You were unfair in giving me a "C," you should have given me an "A."
- I need an "A" in your course or else my grade point average will fall below a "C" or I'll be put on probation.
- It is true that I got a "C+" on the final, but I got a "B−" on the mid-term exam and a "B+" on the term paper, and you may have forgotten that you told me my two oral reports were very well researched and presented. I want to bring my papers back to you for review. Would you be willing to reconsider whether my work has earned me a "B" instead of a "C"?

The first argument is an *ad hominem* attack, placing the blame for the "C" on the bestower rather than the recipient and alienating the target reader. The second argument represents the "me" attitude. It offers a very good reason from the writer's perspective, but it does not address the reader's perspective. Usually, teachers do not give "A's" when they are "needed"; they give them when they are earned. The third argument takes into consideration that the reader is an educator who believes in evaluating students based on the achievements demonstrated in their written work. By providing evidence that is reasonable from the reader's point of view, the student is much more likely to convince the professor that the grade should be reconsidered. The student has also provided the professor with criteria for changing the grade that will be acceptable to them both.

The process of understanding and addressing your reader's perspective when you organize what you write works the same way in business. No matter how complex the relationship of writer and readers, and whether or not you are trying to convince someone to do something or simply dispensing information, *anticipate how the intended audience will react and using your suppositions, write in a way that anticipates and prevents communication interference.*

Since 1973 energy costs for residential home owners have been escalating. One reason for this is that energy-saving window systems have not been incorporated into existing new construction. A new product that could be produced by XYZ Glass Company would be an energy-saving window (ESW). The ESW could be made using current equipment and technology developed over the past 50 years by XYZ.

XYZ currently makes a double-glazed unit of two pieces of glass separated by a dry air space. The performance of this unit for thermal transmittance of heat flow is 0.55 Btu/hr/F. The proposed new product would consist of four pieces of glass separated by three dry air spaces and, also, would incorporate a venetian blind shading device. The thermal transmittance to heat flow, according to research testing, will be 0.18 Btu/hr/F or about one-third of what is presently manufactured. This would save the average homeowner approximately $350 a year in fuel costs based on today's rate. It is estimated that the payback period would average 2.6 years, thus making the ESW an attractive investment.

XYZ's present production facilities would have to be altered only slightly to manufacture this product, and the cost is estimated at $3 million. Marketing indicates a volume sales to $25 million per year for at least the next four years; Production indicates that the product will cost $20 million per year to manufacture.

Figure 4.1   New product proposal.

COMMUNICATION INTERFERENCE

The typical technical communications interference, particularly in writing addressed by lower or middle management personnel to upper management readers, occurs when information on the subject that is of most concern to the reader is available somewhere in the report, rather than featured as the organizing message around which the report is constructed.

In Figure 4.1 is a brief new product proposal written by a product development manager of a glass company. The writer's stated purpose is to write a description of an energy-saving window system that can be incorporated into residential construction. The proposal will be presented to the board of directors of a major glass company. Sprinkled throughout the body of this report are the answers to the central questions that the board of directors would be asking when presented with this proposed new product:

• Yes, there is a viable market—energy-conscious, private home owners, worried by inflationary heating costs.

• Yes, the product can be produced with the company's already existing equipment.

• Yes, the projected payback period makes this procedure an attractive investment.

But instead of organizing this report so that it features these three bottom-line considerations in its opening, the writer begins with a general discussion of

rising energy costs, goes on to a detailed technical description of the new window and ends with the statement that the product will cost $20 million per year
to manufacture. Scattered throughout this presentation are all three of the key
points of concern to readers, but the writer leaves it up to the readers to discover and combine them effectively.

How might the writer have intercepted this communication interference at
the planning stage? There are two procedures for organizing your material to
suit the expectations of your readers—one for short communications such as
letters, memos and brief proposals, and another for long technical reports,
instruction manuals, and so on.

ORGANIZING SHORT TECHNICAL COMMUNICATIONS

If your information is brief enough to be covered in five pages or less, you will
probably want to use the list-making approach to organize material for your
reader. Usually, you will want to make two lists, the Reader-Description List;
and the Content-Description List, and then coordinate them.

The Reader-Description List

First, write down who will read your report (by name and job title). Under the
names and positions of your readers, list all the aspects of your subject that will
be of special concern to them. Try to list these concerns in the form of questions
that your report should answer. Don't be concerned with order at this point,
just list them as you think of them. You may even want to list them as statements
or key words first, and then convert them into questions afterwards.

For the proposal of an Energy-Saving Window, the writer might have produced his reader-description list as follows:

COMPANY BOARD OF DIRECTORS

includes Joel Kane (Personnel Director), Martin Frey (Production Manager), Mary
Eastbrook (Marketing Director), Lewis Payne (Accounting and Budget), and
others.

Market	Is there a viable market?
Production Problems 　labor 　materials 　equipment	Can we manufacture the product with- out prohibitive labor and equipment costs?
R.O.I. (return on investment)	Can we make a profit and how long will it take?
Competition	Do our competitors offer or are they planning to offer anything compa- rable?
Problems/Objections	Are there any reasons why we should hesitate?

If the writer had taken Mathes/Stevenson's suggestion to keep a reader-profile on each member of the Board of Directors, he would have been able to add even more detailed information about the report-reading styles of each person involved. He might have noticed, for example, that "Mary likes the pertinent facts in numbered list form," or "Joel will want to know if the method of producing this window is energy-efficient because he is a fanatic on cutting energy use in manufacturing." Try to include as much detail as you can about your reader(s) in the left column, but also try to convert these details into questions that a written communication can answer.

If you are not certain who will read your report, ask your immediate supervisor or the person or department that requested it. Be polite but persistent. After you have been with your organization for a while, you will get to know the routes of the different kinds of written information you disburse. But there are always exceptions. If you make it clear that you are requesting information about who will read your report and why in order to make the report effective for its users, your superiors will usually cooperate.

Furthermore, in the flurry of multiple assignment-making, administrators often assume that they have given their subordinates more information than they actually have. The burden is on you, the writer, to ask for what you need to know about your readership. Regardless of what others neglected to tell you, the final report will have your name on it. Its success or failure will depend on whether what you have written meets your readers' requirements.

When you have listed all the reader-concerns that you can anticipate and have converted them into a question format, read over your questions. Can you combine any of them into one question? Are one or more of the questions decidedly more important to the specific audience than others? Combine or separate questions where appropriate, and star or underline the priorities.

In the energy-saving window sample, the writer recognizes that the questions regarding potential buyers or markets and potential competitors are both related to the question of whether the product is marketable. He also recognizes that the main question for the Board of Directors is whether or not the Energy-Saving Window will earn sufficient profit quickly enough to make its addition to their existing product line worthwhile.

The Content-Description List

Make this second list right off the top of your head or by glancing briefly at your notes or materials. Don't worry about order or wording, and don't even think about whether this list corresponds to the other list you have just finished. The purpose of this content-description list is to lay out your main facts and ideas so that you can see them all at once in front of you and so that your mind can relate them to each other and to your audience and purpose.

Here is a list that the Energy-Saving Window proposer might have created before writing:

- Since 1973 energy costs for residential homeowners have been rising at an increasing percentage rate every year.
- The windows being developed and installed in new construction are not as energy saving in their design as they might be.
- Because homeowners are sensitive to rising costs of heating and air conditioning, they would provide a broad market for an Energy-Saving Window.
- Our company could produce the Energy-Saving Window using current equipment and technology.
- We could expand on a product already existing—the double-glazed unit.
- The double-glazed unit consists of two pieces of glass separated by a dry air space.
- The Energy-Saving Window would consist of four pieces of glass separated by three dry air spaces, incorporating a venetian-blind shading device.
- It would have a thermal transmittance of 0.18 Btu/hr/F or one third that of present windows.
- Using a window with these properties would save the average homeowner $350 a year.
- The cost of altering production facilities to manufacture the window would be $3 million dollars.
- Marketing projects a volume sales to $25 million per year for at least the next four years.
- The product will cost $20 million per year to manufacture.
- The payback period would average 2.6 years.

Coordinating Your Lists

Once you are certain that you have listed all the points you want to make, place your two lists—reader description and content-description—side by side. First, re-read your reader description to refresh your memory. Then, keeping the picture of who will read your report and why in view, read through your information list. You are attempting to answer two questions:

- Which of these information components should be grouped together and what should each group be headed?
- Given what I know about my readers' needs and expectations, what is the priority message in this list?

In our example, the writer uses his list to identify three groupings into which his information may be divided:

- Marketing
- Production
- Cost and Profit.

These three groupings correspond roughly to the three major questions he decided his readers would want answered. Now he is able to see that given what he knows about the audience and the information from the lists he has just compiled and analyzed, the priority message of his proposal is:

> Our company should add the energy-efficient window to its production line because (a) the market exists, (b) there are no significant competitors, (c) the product can be turned out using existing equipment and technology with only minor adjustments, and (d) the projected payback period would average 2.6 years at a sales rate of up to $25 million per year for the first 4 years.

The writer could use the foregoing statement as the opening of his proposal. Then he would support each of his four reasons with the unifying details.

As you can see, placing all your available information about your readers and your subject matter in front of you facilitates planning. You may be thinking, though, that except for writing down precise figures, there is really no need to list the information point by point for a relatively brief report. Many people who have freshly researched a short project believe they know it well enough to plunge in and write before or without list-making. But even for a memo of less than one page, using this list technique first can improve the effectiveness of the final communication significantly.

Look, for example, at the first version of a memo addressed by the field service administrator of an instruments manufacturer to sales and service personnel on the subject of product warrantee failures (Figure 4.2). From the way it is presented, the memo appears to be a routine description of what a customer should do to repair warranteed equipment that has failed. The intended reader is probably not aware of any special motive the writer might have at this particular time for providing such information. Only toward the end of the memo is the point made rather strongly that the service department is not equipped to handle requests regarding warranty failures and that the service centers repair only warranty equipment; they are not set up to replace warranty equipment.

The field service administrator, who had sent out this memo without getting the desired response, brought it to her company's in-house technical writing seminar and asked the class to tell her what could be done with it to get the priority message across more effectively. She was asked to list all the facts in the memo and read them over to determine what, given her audience of sales and service personnel, the priority message should be.

Only after putting the list in front of her was she able to recognize that the message could be conveyed best in a problem/solution context. The problem was occurring when customers selected the second option—to do the repairs themselves. Because sales and service people were not telling customers what the proper procedures were for making their own repairs on warranty items, the service department and the service centers were besieged by requests they could not fill, and customers were going away confused and dissatisfied. The

---

```
            Sales
To          Sales Secretaries      From   Janis Morgan      Dept./Div. Service/ID
            Service
Date        October 26, 1985                                Location Cleveland
Subject     Warranty Failures                               Copies to List
```

When a customer receives or has warranty equipment that has failed, the customer has several options. The customer will:

1. Contact the nearest service center for repair of the unit.
2. Technically qualify to diagnose and repair the problem.
   a. Contact his salesman.
   b. The salesman will contact the sales correspondent requesting shipment of the replacement items necessary. The shipment is to be made against the original purchase order, with a notation, "credit will be issued upon return of defective items."

This department does not have the facility to handle requests from customers and/or salesmen in shipping parts directly to customers for their warranty failures.

Our service centers do not have the facilities for *replacing* defective warranty items, equipment, and/or parts. Our Service Centers are equipped to *repair* warranty equipment.

JM/mls

Figure 4.2   Memo to sales and service personnel.

solution was to familiarize the appropriate personnel with the proper procedure and encourage them to pass this information on to their customers for everyone's benefit.

Having performed this list analysis with the aid of her fellow writing students, the field service administrator was able to write a second version of the memo (Figure 4.3) which served her purposes more directly and made her message to her intended readers easier to understand and accept.

ORGANIZING LONG TECHNICAL COMMUNICATIONS

The same principle of listing applies when you are organizing for a long report or a technical manual. Once again, you want to put all the material in front of you, so that you can scan and process it into the most appropriate form of presentation. With larger reports, however, you have probably accumulated more information than can be displayed conveniently on a one-page list. Therefore, if you anticipate a technical communication of anywhere from 5 to 5,000 pages,

---

	Sales			
To	Sales Secretaries	From	Janis Morgan	Dept./Div. Service/ID
	Service			
Date	October 26, 1985			Location Cleveland
Subject	Warranty Failures			Copies to List

Background: A customer with warranty equipment that has failed has two
options

1. To contact the nearest service center for repair of the unit
2. To diagnose and repair the problem him/herself.

Problem: A problem arises when the customer chooses option #2 and requests
replacement parts directly from the Service Department. This department does
not have the facilities to handle requests from customers and/or sales
personnel to ship parts directly to customers for their warranty failures.

Solution: We can serve our customers at optimum efficiency, if the following
procedure is used when they request parts for warranty failures that they
intend to rectify themselves:

1. The customer contacts the salesperson with the request.
2. The salesperson then contacts the sales correspondent requesting shipment
   of the necessary replacement items.
3. Shipment is made against the original purchase order, with the notation
   that "credit will be issued on return of defective items".

Compliance with this procedure avoids delay and confusion and promotes
prompt satisfaction of our customers' service requests.

Figure 4.3   Revised memo.

you will want to begin your planning for the writing stage before you begin the
researching, testing, or interviewing that will provide your material.

Basically, you are going to organize for your report as you go along with the
work in your project. When it is time to write up the results, you will have built
a report format into the method that you have used to collect and sort your
information.

To demonstrate how this organize-as-you-go method works, let's begin with
the example of a writer-consultant who has been asked by a government agency
to determine whether gas rationing would be an effective solution to gasoline
shortages in a peacetime economy. First, the writer makes sure he understands
the perspective from which his requesting readers will approach his report. His
reader-description list tells him that the primary concern of the government
agency administrating gas rationing is the feasibility of implementation. They

will want to know what methods have worked best and why. They will also want to know how gas rationing might affect fluctuations in price and whether or not it would create a black market.

Next, *before* the writer begins his research, he creates a structure that will help him prepare his information for the writing of his formal report. This structuring process has five steps:

Step 1: Prepare a master list of topic headings into which you think your subject will divide. Here again, in determining how your subject divides into sub-topics you will keep in mind what you have determined as the special interests of those to whom the final report will be directed. Our sample writer about to study gas-rationing decided that his subject comprised four categories:

- Possible Methods of Distributing Gas Rationing Coupons
- Comparisons of Gas Rationing in World War II and What It Might Be Like in Peacetime Economy of the 80s
- How Gas Rationing Would Affect the Price of Gasoline
- How Gas Rationing Would Affect the Crime Rate

His initial topic heading list reads as follows:

   I. Coupons—distribution of
  II. GR—WW II—cfs.
 III. Price
        A. Up
        B. Down
  IV. GR—crime

Step 2: Label Your Information by Subject-Heading. As you collect notecards, test results, or exhibits, label each in the upper left corner with one of the subject headings on your master list. What you are doing is categorizing your information as you collect it in anticipation of where it will be used in your final report. You are also establishing ways in which information collected from different sources or at different test sites may be connected.

The ideal subject heading for a notecard or exhibit is specific enough to tell you what the material is about but general enough to be used repeatedly to identify and relate different pieces of information by what they have in common.

As you collect your information, you may discover that you need to add topic headings to your master-list. For example, if the writer on gas-rationing discovered valuable information on staffing problems, he might have added a fifth topic heading to his list—Recruiting and Training of Gas Rationing Personnel.

Step 3: Sort Your Information by Subject Heading. When you have completed collecting information and have labeled every diagram, every data sheet, and every quotation or paraphrase from previous literature on the subject with one of the subject-headings on your master list, you are ready to sort your material.

Price – up

*Even with rationing, prices
    rise because*
*(a) if the Arabs resume supply,
    they'll raise prices*
*(b) if they don't, we must open
    up new energy sources, which
    costs money.*
                    Fred Smith, Time, 11/3/73, p 45

Figure 4.4   Sample notecard.

If you were basing your report primarily on material read in published sources, you might have ten sets of notes on each of ten reports. But they would consist of notecards, xeroxes, etc., each labeled in the upper left corner with one of your master headings. In the lower right corner of each card would be the documentation of its source (author, title, date, place of publication, publisher, file and series number, and page reference, or company and date issued. Figure 4.4 provides a sample notecard.

To sort your material for writing, rearrange it, making one pile for each subject heading. The writer on gas-rationing, for example, would have four piles of information in front of him, regardless of how many different sources he used. Each pile would include all the material from each of his sources that concerns one of the four pre-established subject headings on his master list.

Step 4: Re-read and outline the information in each pile to determine the priority message and the supporting evidence under each category, and to determine the overall significance of your research. The outline is designed to serve the same purpose as the content-description list that you use in preparing for short reports. Figure 4.5 is a sample outline of notes under each subject heading for the gas-rationing report. The author's initials in parentheses identify the source of each item.

By re-reading the information under each of his four categories and creating a brief outline, the writer on the feasibility of gas-rationing was able to see fairly easily that each subsection supported the following conclusion: gas rationing is *not* a feasible solution to gasoline shortages in a peacetime economy because it creates more problems that it solves. This conclusion, arrived at by reading sys-

Topic: Is rationing an effective solution to gasoline shortages in a peacetime economy?

I. Coupons—distribution of
   A. To every citizen, driver or not
      1. Each person gets little gas (H.W.)
      2. Coupons sold by nondrivers to drivers (H.W.)
   B. To every car owner
      1. Problem of one-car versus multi-car families (B.J.M.)
      2. Problem of owners holding on to junk cars to get extra coupons (B.J.)
      3. Problem of different cars' mileage per gallon (R.T.)
      4. Problem of different distances to work, shopping, hospitals, schools (R.T.)
   C. To drivers on basis of need
      1. Determined by income level? (J.A.)
      2. Determined by profession? (J.A.)
      3. Able to be regulated efficiently? (J.A.)
II. GR—WW II—Comparisons
   A. Pro
      1. WW II rationing successful (D.W.)
      2. Protected gas supply of vital industries (F.J.)
   B. Con
      1. Patriotic reason to cooperate lacking in peacetime (H.W.)
      2. Defense versus civilian use as basis for distribution no longer applies (H.W.)
III. Price
   A. Up
      1. Determined by conditions outside U.S. (R.S.)
      2. Determined by need to pay higher costs of alternate sources of oil (T.J.)
   B. Down
      1. Supply not exceeded by demand under compulsory rationing (F.J.)
      2. Pressure on refineries reduced (W.T.)
IV. GR—crime
   A. Increase as direct result of GR
      1. Black market created as in Prohibition era (F.J.)
      2. More temptation for private citizens to steal from others' tanks (R.S.)
   B. No increase as direct result of GR
      1. Make coupons legally transferable (F.J.)
      2. Enforce fair distribution with severe penalties (R.S.)

Figure 4.5 Sample outline.

tematically through his pre-sorted information, provides the opening of his report and is supported by the priority messages that the information outlined under each of the four subtopics reveals:

Coupon distribution: No matter which of the three recommended methods is used, equitable distribution cannot be achieved.

GR—WW II—Comparisons: gas rationing was effective in wartime, but, probably, only because of civilians' patriotic motivation to cooperate which is lacking in peacetime.

Price: Some analysts predict that gas rationing will balance supply and demand and keep prices down, but just as many insist that price is determined by conditions outside the United States, such as price fluctuations of foreign suppliers.

Crime: Sources agree that gas rationing will definitely create crime incentives, and there is mixed opinion on whether or not the government's enforcement of strict penalties to deter these types of crime is feasible.

Step 5: Determine the sequence in which you will write up the sections of your outline. The order in which you originally outlined your material need not be the order in which it appears in your final report.

Once you have determined your opening conclusions which will provide the reader with a context for the details, determine the order of presentation of your evidence by putting first the strongest evidence or most important facts necessary to sustain your point. Just make sure that your opening statement summarizes the supporting evidence in the same order in which you will present it in the body of your report.

By using this subject heading/labeling/sorting and outlining technique, you are anticipating the logic of presentation that your final report will require even as you are performing the processing of information that your research requires. Your subject heading list helps you organize your notes and supporting materials; your labeled notes and materials help you formulate an outline; and your outline, written from your notes, leads to a report written from your outline.

This four part system of (1) establishing headings, (2) labeling materials, (3) sorting information, and (4) outlining pre-sorted information, works just as well when you are relying on information gathered from hands-on experience and observing and analyzing your own work environment.

Example: A Report on Cost Efficiency

The superintendent of mine engineering in a copper mining company has received instructions from the mine manager to have his staff examine all mining procedures and equipment for present and potential cost efficiency. The mine manager wants reports on any area of production where the costs might be cut without compromising quality or increasing man-hours. The reports should include an explanation of why the operation could be done less expensively, what changes would be involved, and how to implement them.

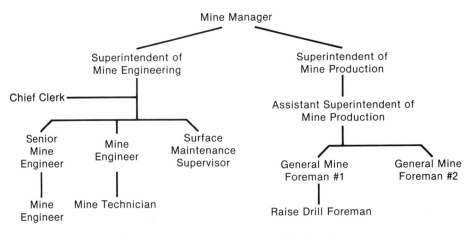

Figure 4.6   Mine company organization chart.

The superintendent of mine engineering relays this directive to the senior mine engineer (see Figure 4.6 for an organization chart) who has had occasion lately to re-examine the company's operation of raise-boring drill cutters. The senior mine engineer believes that the drill cutters could be operated at less cost if there were a mandated uniform operating procedure, which he has designed. He will organize a report using information observed in his own calculations and experiments, conditions of operation observed on the sites involved, and written reports in mining journals and company records which further document his personal observations.

Reader Description

The senior mine engineer's priority audience will be the mine manager, who will make the final decision whether or not to adopt the writer's recommendations. However, initially, his audience will be his individual supervisor, the superintendent of mine engineering. If the superintendent is interested in the report, he may show it to or talk about it with the superintendent of mine production before sending it on to the mine manager. The superintendent of mine production, concerned about how this change in procedure might affect his foremen and the workers they supervise, will probably show the report to his foremen and ask for feedback.

If the superintendent of mine engineering does not consult his peer on the production side *before* submitting the report to top management, the mine manager may ask for feedback from the production crew before deciding on the recommendation. Either way, the mine manager is the senior engineer's main reader, yet he must take into consideration the input of the intermediate readers in determining how to design his report and what to include.

Information Description

How will the senior mining engineer use the recommended organizing system to plan his report? Given his decision that the mine manager is the priority reader, but that the concerns of other subject-area specialists such as the raise drill cutter foremen who will read and react to the report must be taken into consideration, the senior mining engineer sits down to create a list of subject headings for each distinct area of his topic. He brainstorms, trying to anticipate both what an explanation of his subject will require and how his audience needs to be told his information in order to understand and accept it. Here is the list he creates:

Subject-Heading List

I. Analysis of Factors Contributing to the High Cost of Raise Boring
   A. Rock type
   B. Cutter type
   C. RPM
   D. Load per cutter

II. Documentation of the Present High Cost of Raise Boring

III. Possible Methods of Reducing Cost by Altering One or More of the Four Contributing Factors

IV. Implementation Description of Recommended Cost Reduction Method

Using this master list, the senior mine engineer can label each piece of information he collects with the appropriate heading. A previous company report documenting the consistency of rock types in the copper basin might be labeled "I A," for example. A copy of the operating budget submitted by the raise-boring foreman would be labeled "II," and the senior mine engineer's own illustrations of how to graph the load per cutter to achieve maximum load with minimum fatigue would be labeled "IV."

The senior mine engineer would collect all his supporting material, being sure to label each item with one of his four subject headings and creating additional subject headings if he discovers information requiring other categories. When all the information is collected, he will sort it by subject heading and read each pile, outlining the main points as he goes along. Reviewing the outline of his notes, he will determine the priority message of each section and consequently the priority message of his entire report. Analyzing the outline enables him to construct the following opening:

> Raise boring is an extremely expensive mining process. The cost per foot reamed is in the $100–200. range, and cutter operation represents from 25 to 50% of this cost. Of the four factors involved in cutter operation, the first three—rock type, cutter type, and RPM—are not sufficiently controllable to effect a reduction in costs. The fourth factor, however, the load per cutter, can be controlled by the operator to reduce costs. By graphing a cutter load versus penetration rate curve at the beginning of each shift as a standard procedure, the company can save x dollars per year on raise-boring without reducing quality or efficiency.

Having given the mine manager this context, he will proceed through the sections of his outline to support and explain his opening conclusion. First, he will document the present cost of operations (II); then, he will discuss the first three operating factors and why they are not controllable variables (II, A, B, C); then, he will discuss load per cutter and how it might be changed to reduce costs (III). Finally, he will detail the process of graphing the curve and using the resulting charts to operate the drill at maximum penetration rate and minimum cost per foot (IV).

The opening statement will give the mine manager the overall picture he needs to visualize how the senior mine engineer's proposal meets his requirements for a recommendation that cuts operating costs without compromising quality. The supporting sections, each written directly from the writer's outline of his information, provide the details needed by the hands-on production personnel who will be directly affected by the decision to standardize operation procedures.

While this four-step process may seem time consuming, what you are doing is not lengthening your writing time, but rather re-apportioning it and possibly even shortening it. If you go through the suggested planning steps for short and long reports, by the time you get to the actual writing stage, your report will be practically written. Instead of being unable to begin, you will already know how you will begin and also how you will end. You will know the order in which you will present the supporting information, and more importantly, you will know that you have done everything in your power to convey your information to your readers for their intended purposes.

EXERCISES

These exercises are designed to give you practice in decision making and organizing for short and long technical communications.

1. Think of a situation in your school, business, or community in which you would write a memo or letter requesting a change in procedure or an exception to a regulation. For example:

   a. You want to be exempted from a required course or to substitute another course or sequence of courses for the customary one required in your major.

   b. You want to suggest a better way of organizing social or sports activities at your local Y, community center, or place of worship.

   c. You want to recommend a more efficient method of rotating tellers at your bank.

   In order to prepare this memo, make two lists, a reader-description list and an information-description list. Determine your priority message, and write the memo beginning with an opening conclusion and supporting your context statement with an explanation of the facts that appeals to the anticipated concerns of your intended readers.

2. After reading the following report on the food service at a university, identify the priority message and write an outline of the main supporting details. Check to see if there is a sufficient opening conclusion and if the facts are presented logically to the reader?

After completing your outline, make any changes that you believe are necessary to improve the effectiveness of the report. Discuss your changes with those of your fellow students. Revisors should be able to explain how their changes improve the organization of the report.

22 Juniper Dr.
Cleveland, Ohio
December 14, 1980

Marvin Gelb
Director,
Food Service Administration
Smith Commons

Dear Mr. Gelb:

It was brought to my attention that the University Food Service Committee met last night to discuss vegetable preparation and cooking in the University dining commons. My experience as a board plan participant for three years and a food service representative for one year has led me to believe that there is a need for in-depth study of the problems of vegetable cooking for a large number of students. I wondered if you had considered alternative methods of vegetable preparation as a step toward solving this problem. Bob Peterson assured me that you had, but were not sure what type of change was needed.

Because of your sensitivity to the needs of students, I felt that you would be interested in my study. And because of the complaints heard on campus about the Food Service Administration (FSA) and the possibility that your service might be replaced next year, steps must be taken to ensure that students get as much or more from FSA as they would get from other food services. Therefore, I submit the following proposal for your consideration.

PROPOSAL

I have investigated the suitability of various vegetable cooking methods with respect to FSA operations at our university. As the major outcome of my investigation, I will present and evaluate several solutions to the problem of vegetable preparation, allowing you to choose the cooking method(s) most suitable for use in your operation.

PROCEDURE

My investigation of the vegetable preparation problem involved two phases. Phase 1 consisted of consultations with food service personnel, examination of the results of your food service surveys, personal interviews with students, and close examination of the cooking facilities and operations to establish possible solutions to the problem. Phase 2 entailed investigation of the published literature comparing the cooking methods in terms of (1) equipment needs, costs, and ease of use; (2) personnel required; and (3) nutritional detriments and discoloration associated with each method.

ALTERNATIVE SOLUTIONS

The simplest alternative would involve the continuation of present vegetable preparation methods while increasing quality control. The major problem with FSA vegetables is overcooking. Vegetables should be cooked only long enough to make them tender-crisp, in just enough water to prevent scorching. The less water used, the greater the food value retained. The cooking time of carrots, for example, must be controlled within a narrow margin of error. It takes about 13 minutes before thickly sliced carrots begin to approach the tender-crispness desired, but if they are cooked longer than 15 minutes, they start getting mushy. Thus, quick cooking in a minimum amount of water is important for the retention of vitamins, minerals, color, and flavor.

Another problem with vegetables at FSA is their prolonged soaking in steam tables. The pilot light knobs on the steam tables are difficult to manipulate. Without fine adjustment of the flames, it is impossible to control temperature. The flames are meant to keep the vegetables warm, but the "totally on/barely on" nature of the pilot light ends up either cooking the vegetables further (overcooking) or not keeping them warm at all.

Thus, if you choose to continue with the present methods of cooking vegetables, I recommend that you give the pilot lights in the steam tables a tune-up. Also, greater quality control with respect to cooking times and temperatures must be maintained. These improvements would cost nothing and require no additional personnel. However, I wish to propose several alternatives which would produce more marked improvements.

One method would eliminate the need for steam tables altogether. Smaller, more manageable quantities of vegetables would be cooked and served immediately along with the main entrees. The goal is to transfer the vegetables from the cooking kettles to the student's plate in the least time possible. This would require an additional person in the serving line to dish out the vegetables. Your facilities at all three dining commons are equipped with the necessary pots, pans, and small kettles to accomplish this change. The serving lines have room to accommodate the extra staffperson and the vegetables. In fact, they were designed such that no modifications of the fixtures would be needed to make room for the vegetables.

This method would necessitate a reduction in variety and choice of vegetables at each meal so that your cooks could tend to the additional pots and kettles. (The cooks would have to cook periodically throughout the meal, as opposed to the present practice of cooking all the vegeatables at the beginning of the meal.) However, the objective of this method is to serve better-tasting, hot vegetables cooked to perfection—the emphasis here is on a few quality vegetables rather than a variety of mediocre vegetables. Students indicated on your surveys and in my interviews that they usually eat only a few types of vegetables. (They also indicated that they rarely return to the steam tables for seconds, but often return to the serving lines for seconds on main entrees, emphasizing the desirability of serving vegetables on the serving lines.) Therefore, it must be determined which vegetables students like the most.

Students' preferences could be measured objectively by having the serving staffperson keep tabs of what vegetables were served with a simple counter as is used in Carlton Commons on weekends. In addition, students' preferences could be determined subjectively by the dishwashers, who would periodically report to the managers those vegetable dishes that were not being eaten.

Another modification of FSA's present cooking methods would entail cooking the vegetables in a layer of lettuce, eliminating the need for nutrient-absorbing water in the kettles. The lettuce, upon heating, would release enough moisture to steam-cook the vegetables. Again, the vegetables must be cooked only long enough to make them tender. Use of your 40-gallon kettles with the tight-fitting lids, which keep steam and vapor in, would hasten cooking. The vegetables would not lose color, and no additional personnel would be required.

However, layering the kettles with lettuce would demand extra care and preparation time which is not required with the present method of simply dumping the vegetables into the kettles. This type of cooking would require about 25 medium heads of lettuce per week and would cost $10.50 per week if the lettuce were obtained from Hartville Produce at the bulk rate of 22¢ per head.

An even better improvement would be to pressure cook the vegetables. According to an article in *Nutrition Newsletter* (June, 1979), pressure cooking vegetables (just long enough to make them tender) preserves 45% more nutrients than boiling vegetables. Furthermore, this method virtually eliminates discoloration and would require no additional personnel.

To make the best of this method, certain quality controls must be maintained. Only the manufacturer's recommended amount of water must be poured into the cooker. Pressure must be brought up quickly and then reduced to maintain constant pressure. When the vegetables are done, pressure must be reduced as quickly as possible.

Probably the best pressure cooking apparatus for your needs is manufactured by the John Matuszak Mfg. Co. in Columbus, Ohio. Two of their 28 gallon cookers would cost a total of $840, including installation costs. The cookers can easily be tied in to the existing gas lines and would cost 0.46 cents per minute of use or approximately $6.70 per week, a savings of $2.40 over your present cooking methods.

This study provides you with four alternatives to improve your methods of vegetable preparation at the dining commons. When considering this proposal, you will naturally be concerned with the convenience and economic feasibility of these alternatives as well as with meeting the students' needs. Therefore, this proposal is not offered with a "take it or leave it" attitude. The four solutions are not mutually exclusive; instead, they allow you to choose the option(s) which would be most suitable to your needs. My experience and familiarity with the demands of our students convince me that the problem should be solved before the Business Affairs Office reaches a decision on your contract renewal.

CHAPTER 5

The Writing Process: As You Write

The decision-making and organizing functions you perform *during* the writing process are (1) beginning, (2) choosing an order of presentation, (3) making transitions, and (4) closing.

Beginning: If you have followed the organizing procedure described in Chapter 4, you will have an opening conclusion already prepared *before* you write. If not, you may have to begin with a false start, a running commentary that will be transposed and edited after it is written.

Order of Presentation: Recognize which of the three basic structures—description, problem/solution or comparison—best suits your material, and organize your opening conclusion and supporting facts accordingly.

Transitions: Think of a transition from one section or paragraph to the other as an explanation of how two different but related aspects of the same subject are connected. To make a transition, define the relationship: Are they different examples of the same thing? a cause and an effect? a rule and an exception? statistical projections versus on-site inspections? old methods versus new methods?

Closing: Repeat your main point(s) at the end of your report so that your reader is reminded of what the supporting facts signify, and is able to verify that the writer has fulfilled the opening contract.

As you develop your decision-making and organizing skills, your ability to create the appropriate communication structure for each particular job situation will improve. You will be able to prepare and analyze lists and outlines that will lead you efficiently to the composing stage of the technical writing process.

This chapter addresses four questions about writing effectiveness that usually arise while you are composing:

- How do you get started?
- How do you determine an order of presentation?

- How do you make smooth and clear transitions from paragraph to paragraph and from section to section?
- What do you say at the end of your report, especially if you began it with your conclusions?

GETTING STARTED: PLANNED VERSUS FALSE STARTS

Ideally, you have gone through the pre-writing organizational procedures recommended in Chapter 4 for preparing short or long technical reports. If so, your initial panic over what to put first on that intimidating blank page is avoided because you know what to put first—the opening conclusion that you determined by analyzing your information for a priority message.

The Planned Opening

Formulating an opening conclusion becomes fairly simple when you have the list of your information (for short reports) or the outline of your information (for long reports) in front of you, and you have already analyzed that list or outline in conjunction with your reader-description list to determine the central message and the main supporting points. Instead of panicking or guessing, you simply transpose this information to write the opening summary of your report.

In Figure 5.1, for example, is the opening summary of a report written by a

Figure 5.1 Materials characterization of commercial and experimental lamp lead wires: opening summary.

A thorough materials characterization was performed on proposed in-house Process 3 to disprove patent encroachment on the currently available commercial Product A. Characterization on as drawn and annealed samples to 1000°C in finished wire sizes was done by

1. Tensile testing
2. Bend testing (Tinneous-Olson)
3. Resistance measurements
4. Microstructural analysis (optical, and scanning and transmission electron microscopy).

Results showed mechanical and structural differences sufficient to qualify Process 3 material as a new product. First, Process 3 material has a lower strength and bending stiffness than Product A (although both are sufficient for its intended application). Second, particle sizes are two orders of magnitude larger in Process 3; and third, location is intragranular rather than dispersed throughout the crystallites as in Product A.

The complete process description for patent disclosure is being reviewed by the Legal Department and will be distributed following approval.

metallurgical engineer, titled "Materials Characterization of Commercial and Experimental Lamp Lead Wires." If you divide this opening summary into its components, you find:

Purpose: what was done and why (sentence 1)

Method: what it was done to and how (sentence 2)

Results and Their Significance: what happened and what it means (sentence 3)

Supporting Evidence Summarized: what the conclusions are based on (sentences 4–6)

Future Implications: what will happen next (sentence 7).

The writer was able to determine that these were the necessary components of the opening by looking at the subject headings into which she had divided her information while collecting it:

- Definition of Problem to be Solved
- Materials Specifications
- Methods Used
- Results of Each Test
- Implications
- Present
- Future

In other words, the proper classification and division of your information in the planning stages enables you to formulate the components of your opening as you go along.

The False Start

You may not be a born list maker. Suppose you had done the work for the materials characterization report without reading Chapter 4 or without having had time to practice the listing procedure. You have in front of you tables and graphs documenting the results of tests performed to determine the structural and functional differences between commercial Product A and a propsed in-house substitute, Process 3 material. The information includes tests made on four other experimental processes in addition to Process 3, which were also considered as possible replacements for Product A.

You also have on your desk two memos from your supervisor: one requests a report documenting the choice of Process 3 over the other possible process replacement materials; and the other requests a materials characterization report on Process 3 for purposes of patent disclosure. If you have not planned the collection of your materials in anticipation of complying with these report requests, what can you do to get started?

Begin with a false start.

Instead of staring helplessly at an empty page or screen, start to recount for

Lamp lead material, at present, is available through a single supplier. The obvious drawbacks in the absence of competition are lack of price controls and material availability. There is also little incentive to tailor products to special needs of a customer.

In response to a request by the Lamp Products Department to find an alternate source for lamp lead material, a process for making a suitable copper composite material in-house was invented by the Lamp Metals Laboratory.

Several methods of producing a material suitable for lamp leads, similar to Product A, were tested in the laboratory. Each product was evaluated by mechanical and physical properties, and consistency from batch to batch. As a result, Process 3 was chosen as a test method. A thorough materials characterization was then necessary to disprove patent encroachment by showing differences from the commercially available material, Product A, which is now used for lamp leads. The in-house process is explained in detail in a patent disclosure, now being reviewed by the Legal Department. The methods by which material properties were compared are:

1) Tensile testing

2) Bend testing by Tinneous-Olson method

3) Resistance measurements

4) Microstructural analysis by three methods: Optical micrography, scanning (SEM) and transmission (TEM) electron microscopy.

Although tests indicate our material is not as strong as Product A, specific requirements for applications of the product are met by Process 3 material. Choice of Process 3 over the other processes was made based upon three considerations: 1) this process produced a stronger material, 2) material properties were very repeatable from batch to batch, and 3) optical structural analysis by optical microscopy, SEM, and TEM techniques confirmed findings. These comparisons were conducted on drawn wire samples before and after annealing, in the sizes used in manufacturing.

Figure 5.2   Writing to get into your subject.

yourself in writing what happened and what you did. You are writing to get into your subject, which accomplishes two purposes:

1. It loosens you up and gives you confidence by letting you see that you can get started.
2. As you write the play-by-play account of the subject from *your* point of view, you begin to observe what the readers need to know to understand it from *their* point of view.

Our sample report writer (Figure 5.2) starts with the reason for initiating the new product search, then gives a description of what was done to develop a new product, and of how Product 3 was chosen over the other possibilities. She describes the methods of testing and the results. As she re-reads this version of the material, organized from her point of view, she recognizes (a) which details are either irrelevant to the patent disclosure, or (b) should be relegated to a background section, and (c) what part of this material is relevant to her readers.

This type of revision is less efficient than organizaing beforehand while you are first accumulating your information. But if formal methods of organizing really intimidate you, plunging right in may be your best choice, for a while. Just remember that your first paragraph or first page will probably be discarded or relegated to another section of your report. It serves as a way for *you* to enter your subject, but, usually, it does not serve as a way for *your reader* to enter it.

Your rough draft is really a list or outline of your topic in paragraph form. It takes longer to prepare than a list or outline, but when used constructively, it enables you to have a body of material fully in view in order to restructure it for the intended audience. Then, too, having to write two versions of a report when you could have written just one may provide an added incentive for becoming a list maker.

RECOGNIZING AND CREATING STRUCTURES

Familiarize yourself with the basic structures of a written communication and learn to recognize which is best suited to your information, audience, and purpose. Most technical communications use the following structural patterns singly or in combination:

- Description
- Problem/Solution
- Comparison

The Descriptive Structure

Your material should be presented using the descriptive pattern if your primary goal is to explain a situation, condition, object, or process. If you select the descriptive structure, your opening conclusion should include in the following order

- The reason why the reader needs or would want to understand this description
- A summary of the main features and why they are relevant to the target readers.

The materials characterization report discussed earlier, for example, would be suitable for a descriptive format, if the writer's main purpose were to explain either to upper management or to hands-on technicians how materials are characterized for the purpose of patent disclosure. Such a report addressed to upper management might begin as follows:

> In order to obtain a patent clearance for proposed Process 3 material, it is necessary to prove that Process 3 material differs functionally and structurally from

HOW TO CONTROL FOOD COSTS IN THE RESTAURANT BUSINESS

In the majority of restaurant operations throughout the country, food costs represent the single largest daily expenditure. The success or failure of a restaurant operation is often dependent on management's ability to control these costs effectively.

Effective controls are influenced by two special features of the restaurant business: (1) although the restaurant is essentially a manufacturing operation, its product is highly perishable; and (2) a large quantity of food items must be prepared for each daily meal period.

If food costs are to be controlled, the restaurant operator must:

1. *Accurately predict what the customer is going to buy.* The closer an operator can come to 100% accuracy, the more control he will have over his products. By knowing in advance which items are more likely to be sold, he/she can prepare a forecast of costs. This forecast can then be compared (on a daily or weekly basis) to the actual costs to see if the actual costs are in line with an established profit plan.

2. *Purchase raw material in accordance with the predictions.* With an accurate prediction, it is easier to purchase only those estimated quantities. Such practices as underbuying or overbuying can be greatly reduced.

3. *Eliminate waste and loss from the time foods are purchased until they are sold.* Wasting good food instead of using it properly can add considerably to the actual food cost. Food costs will be minimized by closely controlling the movement of merchandise from the time of purchase to the time of sale.

4. *Skillfully prepare and portion the product to avoid excessive costs.* By adopting standardized portions, an operator can keep a very accurate record of what his costs should be. If actual costs begin to get out of line, this trend can be spotted and corrected more easily.

Figure 5.3   A description structure.

Product A sufficiently to constitute a "new product." For this purpose a materials characterization study was performed on Process 3 material.

This would be followed by a summary of each test and each result. The body of the report would give test conditions, specifications and results in detail.

A second example of a description structure is the opening of a report written by a restaurant manager in Marietta, Georgia. The intended audience is new managers being trained in restaurant operations (Figure 5.3).

The opening conclusion is followed by four sections, each of which outlines in detail the four operations summarized in the opening.

The Problem/Solution Structure

This format, also described as the question/answer or cause/effect structure, is useful if you are a financial analyst, a quality control engineer, a research and

Problem: At present, lamp lead material, is available through one supplier only. Absence of competition results in no price controls, unreliable delivery, lack of cooperation in suiting the product to the special needs of each customer.

Solution: The Lamp Metals Laboratory has invented a process for making a suitable copper composite material in-house. This Process 3 material was tested against five other possibilities. Each was evaluated for mechanical and physical properties and for consistency from batch to batch. Process 3 proved superior to the other five in each test.

Materials characterization tests have been performed on Process 3 and Product A and sufficient differences have been documented to warrant patent disclosure.

The full disclosure will be released after review and clearance by our legal department. It appears, however, that producing Process 3 material in-house will solve the problems created by being dependent on the supplier of Product A.

Figure 5.4   A problem/solution structure: engineering.

development scientist, a systems analyst, or anyone whose profession involves finding answers to questions. Your material should be presented in a *problem/solution structure* if your primary goal is *to explain a negative condition and recommend a positive way of dealing with it.*

If you select the problem/solution format, your opening conclusion should include (in this order):

- A clear statement of the problem
- A recommended solution
- A summary of the reasons for proposing this solution and of the methods used to arrive at this decision
- A summary of future implications, if applicable.

If the primary purpose of the materials characterization report were to recommend Process 3 material as a suitable replacement for the unsatisfactory Product A, the opening conclusion might look like Figure 5.4. This opening would be followed by sections on the background of the problem, the test results for each material considered, and the meeting of requirements for patent disclosure.

Another example of a Problem/Solution structure is the opening summary for a proposal written by an undergraduate accounting major. The writer has examined a firm's cash flow problem and is proposing that they hire him to provide the solution (Figure 5.5).

The Comparison Structure

Sometimes making comparisons for your reader is your primary concern: to report test results to a supervisor; to convince a potential customer that one

---

Thorough examination of the financial statements and records of KIZ Inc., and interviews with company officers indicate that your firm is experiencing a significant problem in cash management.

Because KIZ Inc. has a) high inventory turnover and b) seasonal sales, it requires high working capital part of the year and minimal working capital at other times in order to operate efficiently.

To solve this problem, AWE & AWE proposes a two-phase plan that will

1. rectify the immediate lack of working capital by the preparation of a working capital pro forma statement

2. initiate measures to forestall future working capital shortages by compiling a working capital projection through 1986.

Figure 5.5   A problem/solution structure: accounting.

product is better than another; to aid a surgeon in determining which procedure to use and when. At the time, the comparison is made not for its own sake, but in support of what is primarily a Problem/Solution format.

Both materials characterization reports, the one intended to demonstrate why Process 3 is better than the other five materials considered, and the one that compares Process 3 to Product A for patent disclosure, are written in a comparison structure. In each case, however, the comparisons are used to support a problem/solution format.

If the primary goal of your report is to show the similarities and differences between two or more objects, processes, concepts, or career choices, you will want to package your information in a Comparison Structure. Using this format, your opening conclusion should contain in this order:

- A summary of your conclusions
- A summary of the most significant similarities and differences on which you base your conclusions
- A statement, if appropriate, on the present or future implications of the comparison's results.

If the main purpose of the materials characterization report were to compare Process 3 material to Product A, the opening might read as follows:

Tested for strength, consistency from batch to batch and microstructure, Process 3 proved different from Product A in ways sufficient to enable patent disclosures but insufficient to decrease the effectiveness in the desired application. Particle size is larger in Process 3 material and its location is intragranular rather than dispersed throughout the crystallites as in Product A. As soon as patent approval is granted from our legal department, a step up of the product to production-sized batches should take place. Although tests indicate that Process 3 material has a lower strength and bending stiffness than Product A, it produces equally effective lead wires under comparable conditions.

Points of Comparison	Type A	Type B	Type C
1. Cost			
2. Style			
3. Longevity			
4. Specifications			
5. Standard equipment			
6. Optional equipment			
7. Load capacity			
8. Size			
9. Safety features			
10. Mileage			
11. Trade-in-value			

Figure 5.6   A comparison frame.

When you suspect that your report will fit best into a Comparison format, you may want to do your beforehand listing or outlining using a comparison frame.

The Comparison Frame

A comparison frame is a table in which the points of comparison (criteria used to evaluate items compared) are listed vertically at the far left and each item to be compared is listed horizontally across the top. If, for example, you were comparing two kinds of cars, you would list all of the categories to be considered in evaluating the two under points of comparison. Then, you would be able to fill in the information for each criterion under each car (Figure 5.6).

This method puts in front of you all the information you need to formulate an opening, clearly and concisely. It allows you to evaluate what the information signifies to you and your readers.

Figure 5.7, for instance, was prepared for a health insurance agency that was re-evaluating its primary collection procedures. Their choices are in-house centralization of the direct premium-collection process, or centralization of the collection process in a financial institution. A quick look at this figure shows even the outsider that the basic difference between the two choices under each of the first four criteria is time expended in processing. Based on criterion 1, handling of deposits, the financial institution can function more quickly; but under criteria 2, 3, and 4, in-house centralization provides immediate resolution, whereas centralization in a financial institution requires re-routing delays.

Items 5 and 6 involve availability and cost of equipment and personnel, and in both cases an in-house operation would cost less.

In other words, the comparison frame tells the writer/analyst that in five out of the six criteria, in-house centralization would be less expensive and more efficient. The only disadvantage to in-house centralization is indirect deposit of premiums. If this is compensated for by the savings realized under each of the other five criteria, in-house centralization of premium collection is the preferable choice.

Points of Comparison	Collection In-House	Collection by Financial Institution
Handling of deposits	indirect via Federal Reserve Bank to us—deposit slips returned two days later	direct: deposit slips returned daily
Processing of exceptions	immediate	would have to be returned
Processing of out-of-balances	immediate resolution	would have to be returned to us for resolution
Processing of mutilated bills	immediate	would have to be sent back to us first
Availability of microfilm	individual access	at additional cost only
Staffing requirements	can utilize employees already on hand	would need to hire and train additional employees

Figure 5.7   Comparison table: collection procedures.

In-House	Financial Institution
Advantages	Advantages
1. Employees already on-hand used.	1. Deposit would be made to Trust Fund daily.
2. Mutilated bills can be searched and/or prepared immediately.	2. Deposit slips would be returned daily.
3. Exceptions can be searched immediately.	Disadvantages
4. Deposits can be made daily.	1. Special employees would have to be hired.
5. Out-of-balances can be resolved immediately.	2. Mutilated bills would have to be sent to us.
6. Microfilm would be available immediately.	3. Exceptions would have to be sent to us.
Disadvantages	4. Out-of-balances would have to be resolved by us.
1. Deposits would be made to Trust Fund via Federal Reserve Bank.	5. If out-of-balance, deposit would not be made.
2. Deposit slips would be returned via Federal Reserve Bank—2 days.	6. Microfilm would only be available at additional cost.

Figure 5.8   Collection procedures: advantages and disadvantages.

Instead of making a comparison frame, the writer might have listed the advantages and disadvantages under each choice (Figure 5.8), but note how much repetition would be involved, and the author would still have to do the time-consuming analysis of the implications of these advantages and disadvantages, which the comparison table has already made apparent.

The opening summary based on the comparison table in Figure 5.7 might read as follows:

> In-house centralization of premium collection is more time and cost efficient than centralization of premiums collection in a financial institution. In-house centralization may be implemented with existing staff and microfilm equipment. It allows immediate resolution of problems arising from out-of-balances, exceptions, and mutilated bills. Although centralization in a financial institution would speed up deposit-slip return, this advantage does not compensate for the additional costs incurred by delayed resolution of routine problems and the necessity of hiring and training new personnel and paying extra for microfilm privileges.

This opening summary would be followed by the details of each criterion compared, keeping to the same order in which they appear in the introduction and using the material from the comparison table.

ORDERING THE SECTIONS OF A REPORT

If you choose one of the three standard formats—Description, Problem/Solution or Comparison, your opening conclusion tells you and your reader what the sections of your report are and in what order they should appear.

In the case of the Description format selected for the report on how to control food costs for restaurant managers, the opening summary indicates that you will have a section on each of the four procedures introduced:

1. How to accurately predict what the customer is going to buy
2. How to purchase raw material based on your predictions
3. How to eliminate waste and loss from the time foods are purchased to the time they are sold
4. How to skillfully prepare and portion the product to avoid excessive costs.

In this instance, the four procedures for food cost control are presented in an easy-to-follow chronological sequence: what to do before, during, and after purchase.

If you are using the Problem/Solution format, your opening will use in miniature the same sequence that you will follow in the body. In the opening you stated the problem concisely: in the body, you will fill in the background necessary for the reader to understand the problem, its causes, its implications. In the opening you stated the solution concisely and summarized your reasons and the methods you used to reach your conclusions: in the body, you will explain in detail each of your reasons for the proposed solution and each of the tests, conjectures, inspections, interviews, and so forth that led you to that solution.

For example, in the version of the materials characterization report that uses a Problem/Solution format, the body would cover topics in a sequence like this:

- Background
  - What lamp lead material is used for
  - How it fits into the production sequence
  - Problems arising from having only one supplier
    - No price controls
    - Unreliable delivery
    - Product adaptability to special needs unavailable
  - Past history of attempts to deal with these problems and why they were unsuccessful
  - Lamp Production Department's request to Lamp Lab and why it was prompted at this particular time
- Materials and Methods
  - Procedures used by Lamp Metals Lab to select and develop new materials for testing
  - Materials chosen (i.e., Process 3 and five others)
  - Testing methods, equipment, and conditions of testing described thoroughly
  - Test results and interpretation
- Results
  - Reiteration of reasons for selecting copper composite material of Process 3
  - Reiteration of reasons for believing Process 3 material is suitable for patent disclosure.

If you use the opening suitable for a Comparison format, it tells you and your reader what order of presentation the body of your report will follow. The report recommending in-house premium collection, for instance, tells you at the outset that the six criteria on the basis of which the decision was made should be discussed in detail in the following order:

1. staffing
2. microfilms
3. out-of-balance
4. exceptions
5. mutilated bills
6. deposit slip return

Why in this order? Why not detail the information on each criterion in the order in which each appears in your comparison frame? You want to present your detailed comparisons in the same order in which you have grouped them in the opening for two reasons: so that you will write consistently and clearly, and so that your reader will follow your line of reasoning easily.

Think of your opening as a contract that you make with your reader. You promise in the beginning of your report to discuss a given subject from a given vantage point and to express and analyze observations and conclusions in a stated order. You make commitments that readers will expect you to fulfill. If you name five advantages and one disadvantage to centralized in-house premium processing, then your reader expects to see five advantages and one disadvantage discussed in the body of your report in the same order in which they appear in the opening. If you claim that those advantages outweigh the disadvantages, the reader expects to see evidence supporting that statement. If, instead of taking a position on in-house versus external centralization, you state that your report will simply explain the advantages and disadvantages of each and let the reader decide, then that is the contract your report is expected to fulfill. Otherwise, readers' focus shifts from the information to concern with where they lost you.

MAKING TRANSITIONS

Once you understand the order of presentation that is dictated to you by the format you select and announce in your opening statement, the division of your material into sections and paragraphs will follow nicely. Questions will still arise, however, about how to get from one section or paragraph to the next. These leaps of logic require "bridges" or transitions that are sometimes difficult for you and your reader to make smoothly.

Whenever you are having difficulty moving from one section to the next, one topic to the next, or explaining the connection between two related but different aspects of the same topic, try to define the nature of the relationship. Are the topics or sub-topics

- Different examples of the same thing?
- A cause and an effect?
- A rule and an exception?
- An old procedure and a new one?
- Tests based on physical observation versus tests based on simulated conditions?
- Statistical projections versus on-site inspections?

You must determine the relationship between what you have just written and what you are going to write before you can make that relationship clear and acceptable to your readers.

If you were writing a descriptive report on smoke detecters, for instance, it would help your reader move from one description to another if you used a transitional sentence summarizing how A and B are alike and how they are different:

> The ionization type is just as effective as the photoelectric smoke detector. It differs only in the *way* it detects fires. The photocell detects light frequencies, while the ionization detectors register particles of air.

Actually, each section or paragraph of your report is a report in miniature. If you begin each with an opening summary, the transition from part to part follows the logic dictated in the opening. In the food purchasing section of the report on controlling food costs, for example, the writer indicates at the beginning that there are two efficient buying methods that keep cost to a minimum: competitive buying, and quantity and specials buying. He then discusses one after the other.

Whenever possible, give a collective name and a fixed number for items to be discussed in a particular section of your report: For instance, there are six reasons, three advantages, four tests, or five variables. First name them all, and then discuss each in detail in the order named. Defining the relationship of the items to be discussed and giving the number of items at the beginning makes transitions from point to point clearer and easier for you and for your readers.

In the section on "Cost Control Procedures" there are several examples of how defining the relationship of a group of facts and establishing their number at the outset simplifies the movement of the discussion from one point to another. The writer is establishing a context in which the reader can grasp the flow of information. Study the excerpt from the Food Cost Control Report in Appendix A, in which key grouping, defining, and numbering sentences that aid transitions from part to part have been underlined to explore how transitional sentences improve and shorten reader-processing.

COMING TO A CONCLUSION

Most people will agree that the only writing task more difficult than getting a report started is the task of deciding how to end it. Whether or not you use the technique of opening with a concluding summary, you may have found that when you finish the body of your report, you feel as if you have nothing else to say. You will have to either repeat yourself or else launch into grand cliches.

It may help you to keep in mind that the conclusion serves two purposes for your readers:

- It recalls for them what the terms of the contract were as announced in your opening.
- It confirms for them that the expectations created by the opening contract have been met.

Properly used, your closing statements are not redundancies: they are useful reminders.

Appendix B contains a NASA report on parawing gliders that exemplifies the effective use of repetition in the ending of a report. Notice that the section headed "Concluding Remarks" repeats, sometimes using the same phrasing, the major points made in the opening sections labeled "Summary" and "Introduction." Even though the report is only a few pages long, it takes readers through detailed technical material, graphs, and diagrams. At the end, they need to be reminded of the context in which they are examining this presentation of facts.

Another practical fact to consider is that many readers do not read a report in its entirety or in the order in which it was written. A reader may read the conclusion first, not realizing that you have provided an opening summary. Or, a reader may read only the section labeled "Conclusion." Because different readers will read your report for different purposes and because many do not read it from cover to cover, the central messages must appear in more than one place and often be repeated at the end. Readers should conclude a report feeling confident that it has told them what they wanted to know, or, at least, what it promised to tell them at the beginning.

EXERCISES

These exercises are designed to give you practice in dealing with each of the 5 questions discussed in this chapter that come up as you write:

1. Write a running commentary, in the exact order that the events occurred, of what you did to complete a recent school or work project. Then rewrite (repackage) the information so that it delivers a priority message to the reader and fits one of the three recommended structures: Description, Problem/Solution, or Comparison.

2. Select two technical report samples from this book, from a professional journal, or from a library collection. Read the reports to identify transitional passages and places where necessary transitions may have been omitted. Consider which transitions work well and which do not and why. If you think a transition is missing, try writing one that would suffice.

3. Select five report samples from those available throughout this book or from technical reports available in your school or business library, or from your instructor. Read the reports and determine in each case what the conclusion contributes to the report structure. If you feel the conclusion is ineffective or incomplete, rewrite it and explain what you changed and why.

The Writing Process: Using Visual Aids

While you are structuring and writing a report, you will be making three decisions about visual aids:

• When to use illustrations
• Where to place them in your text
• Which of the many types of illustrations to use when.

If you needed to visualize something in order to write about it, chances are your readers will need to see it, too. Put all illustrative material next to or opposite the written passage that refers to it. Learn the basic strengths and weaknesses of graphs, tables, diagrams, illustrations, and photographs, so that you will have the criteria for identifying the most appropriate visuals for each of your purposes.

WHEN TO USE VISUAL AIDS

Since many experts in technology tend to be visually oriented, illustrations often provide the core or focus of a technical communication. Experts frequently look at diagrams first, and consult the text afterwards, if there is something they still need to understand. On the other hand, if you are far more familiar with the object, process, or location described than your readers, they will need a visual aid, even if you do not. If your discussion involves comprehending quantities or percentages in relation to one another, a table or graph is probably more effective than long paragraphs of data. Given the increasing capabilities of computer graphics, technical writers are relying more and more on good illustration. In the report on Food Cost Control in Appendix A explanations are clarified by illustrative tables. To explain how to determine the standard menu item, the writer had to look at or at least picture in his mind an actual yield test to recall how it is done, what it indicates, and how it is used. Therefore, in writing the

Standard Yields

Yield means the net weight or volume of a food item after it has been processed and made ready for sale to the guest. Standard yields are those that result when an item is processed according to established standard preparation procedures.

Yields serve two major functions. First, they are important in determining the selling price of an item. Second, they provide management with a potential dollar amount of income per item based upon how many portions (yield) can be carved from that item.

Item: Beef Rib, Oven Prepared	Grade: Choice
Item Cost: $28.28 at $1.40/lb.	Weight: 20 lb., 4 oz.

Summary of Yield Test

Cooking and Carving Details	Weight	Percentage of Original Weight	Cost per Servable Pound
Servable Weight	11 lb. 3 oz.	55.2	$2.53
Loss in Carving	5 lb. 3 oz.	25.6	
Loss in Cooking	3 lb. 14 oz.	19.2	
	20 lb. 4 oz.	100.0	$1.40

Figure 6.1   Excerpt from report on *Food Cost Control* (Appendix A).

description, he included a sample table of a yield test so that his readers could look at it as they read his explanation.

Notice the order of presentation (Figure 6.1):

• The term "standard yield" is defined (paragraph 1, sentences 1 and 2).
• The two major functions of "standard yield" are described (paragraph 2, sentences 1 and 2).
• A sample yield table is presented with an explanatory note.

As with the table just illustrated, when your reader needs to look back and forth between the writing and the graphics, place them next to each other on the same page, or on facing pages to facilitate the reader's task.

DECIDING WHAT KIND OF VISUAL AIDS TO USE

To familiarize yourself with the basic types of visual aids—charts, tables, diagrams, illustrations, photographs, blueprints—note good and bad examples in the textbooks and professional journals that you use in your class or on your job. When you see a diagram that really helps you understand the writer's point, ask yourself why it is effective:

• How is it placed in relation to the written section that refers to it?
• How are the parts labeled?
• Why is it easy to interpret?

When you find a visual that is confusing or seems unrelated to the written section that refers to it, also ask yourself why. When you read a technical passage that makes you wish you had an illustration, try to decide what type of aid you would need. By learning what you as a reader require in the way of visual aids to help you understand technical subjects, you will begin to understand what your readers need and how to provide it for them.

The three most frequently used visuals are tables, graphs/charts, and diagrams/illustrations. In the following sections, each type is described and exemplified, and guidelines are suggested for its use.

Tables

Tables are used most often to facilitate the viewing and comparison of numbers. The *strength* of a table is that it places numbers or data before the reader in a compact format, making them easy to analyze. The *weakness* of a table is its limited ability to stress any particular relationship between the numbers presented. To express one or more connections, you will need to supplement or replace a table with one or more graphs (Figure 6.2).

Most tables consist of a *title,* a *box head* (sub-titles horizontally across the top), a *stub* (sub-titles placed vertically to the left) and a *field* divided into columns. (see Figure 6.3).

Title
As J. S. Hanna observes, "a table title ought to do two things only—and each briefly": "summarize table contents"; "distinguish those contents from the similar contents of like tables."[1] For example, the title "Oxidant Trends in the South Coast Air Basin" is too general; the title, "Oxidant Trends in the South Coast Air Basin/Three-Month Averages of Daily Maximum One-Hour Oxidant Concentrations for July, August and September (pphm)" is too long and too complicated. However, "Highest One-Hour Oxidant Concentration in Parts per Hundred Million" is a title that combines brevity with thoroughness.

Box Head and Stub
The selection of horizontal and vertical headings depends on deciding beforehand what you want to compare to what. The table in Figure 6.3 for example, is not really the most effective presentation of data because it reports data by type (worst case, smog season, yearly), while the author's real intention is to compare cities.

Field
A table serves its purpose only if its columns and figures are easy to read. To ensure readability,

- Avoid crowding
- Leave adequate space between columns

[1](J. S. Hanna, "Six Starts Toward Better Charts," *Technical Communication*, 29 Summer, 1982, p.4).

Carson Pirie Scott & Co.

Earnings by fiscal quarter are summarized in the following table
(amounts in thousands:)

Earnings

| | 1980 | | 1979 | | 1978 | |
Quarter	Amount	Per Share	Amount	Per Share	Amount	Per Share
1st........	$ 544	$ .16	$ 1,355	$ .39	$ 1,174	$ .34
2nd	818	.23	2.070	.60	2,246	.64
3rd	2,081	.60	3,688	1.07	3,545	1.02
4th	8,600	2.47	5,495	1.58	5,241	1.50
	$12,043	$3.46	$12,608	$3.64	$12,206	$3.50

Growth in Earnings & Dividends per Common Share

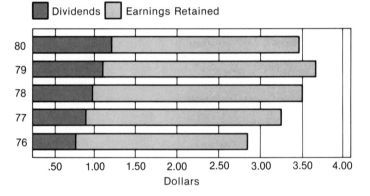

Figure 6.2 A table and a graph using the same data for different purposes. (From the
Carson Pirie Scott & Co. annual report, 1980.) The table lists the earnings per quarter
for 3 years; the chart uses these figures over a five-year period to show dividends and
earnings retained in 1980 compared with previous performance.

- If table will be reduced, allow for effects of size change on visibility
- Make sure the scale of your drawing is appropriate to the typeface that will
  be used.

A good example of a table design is Figure 6.4; a bad example is Figure 6.5.

Graphs/Charts

The terms "graph" and "chart" are used almost interchangeably to describe the
same visuals. As opposed to tables, graphs/charts illustrate selected points about
facts rather than just the facts themselves.

Title → Highest One-Hour Oxidant Concentration
in Parts per Hundred Million

Station	Average Amount by Year					
	1963	1964	1965	1970	1971	1972
3 Highest: July–September						
Anaheim	22.3	20.0	33.7	29.7	27.3	28.7
Azusa	32.3	38.7	44.0	52.0	40.3	36.0
Burbank	32.7	30.7	35.0	31.7	27.3	25.0
Pasadena	36.7	36.0	36.7	45.3	39.7	32.0
Daily: July–September						
Anaheim	11.4	9.6	15.9	10.7	8.9	8.7
Azusa	19.8	24.2	24.4	28.8	22.9	18.1
Burbank	15.1	15.6	20.9	18.5	16.1	13.2
Pasadena	20.0	21.9	21.6	25.7	20.6	17.1
Daily Annual						
Anaheim	8.0	7.3	11.1	7.6	6.3	5.9
Azusa	13.2	14.3	15.3	16.0	13.1	12.0
Burbank	8.9	8.5	12.2	10.9	9.3	8.6
Pasadena	13.2	13.0	13.7	15.0	12.0	11.1

(Box head ← , Stub → , Field ←)

Figure 6.3 Sample table title and format. (From J. S. Hanna, "Six Starts Toward Better Charts," *Technical Communication*, 27, Summer, 1982, p. 5.)

Figure 6.4 A table with a readable field. (From J. S. Hanna, 1982.)

Table 2
OXIDANT TRENDS IN THE SOUTH COAST AIR BASIN
Three-month Averages of Daily Maximum One-hour Oxidant Concentrations for July, August and September (pphm)

Station	1963	1964	1965	1966	1967	1968	1969	1970	1971	1972
Anaheim	11.4	9.6	15.9	14.1	12.5	11.9	13.7	10.7	8.9	8.7
Azusa	19.8	24.2	24.4	25.8	26.8	21.9	28.0	28.8	22.9	18.1
Burbank	15.1	15.6	20.9	17.0	22.5	19.0	19.4	18.5	16.1	13.2
Pasadena	20.0	21.9	21.6	22.2	22.6	22.3	27.4	25.7	20.6	17.1

Table 4

OXIDANT TRENDS IN THE SOUTH COAST
AIR BASIN

Average of Three Highest One-hour
Oxidant Concentrations for July, August
and September (pphm)

Station	1963	1964	1965	1966	1970	1971	1972
Anaheim	22.3	20.0	33.7	30.3	27.7	27.3	28.7
Azusa	32.3	38.7	44.0	41.7	52.0	40.3	36.0
Burbank	32.7	30.7	35.0	28.0	31.7	27.3	25.0
Pasadena	36.7	36.0	36.7	39.0	45.3	39.7	32.0

Figure 6.5   A table with a field that is difficult to read. (From J. S. Hanna, 1982.)

Figure 6.6   Eight types of bar charts. (From Calvin F. and Stanton E. Schmid, *Handbook of Graphic Presentation,* 2nd ed., New York: John Wiley and Sons, Inc., 1979, pp. 61–62.)

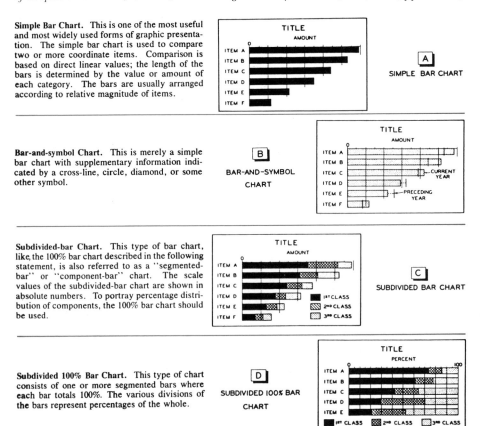

Simple Bar Chart. This is one of the most useful and most widely used forms of graphic presentation. The simple bar chart is used to compare two or more coordinate items. Comparison is based on direct linear values; the length of the bars is determined by the value or amount of each category. The bars are usually arranged according to relative magnitude of items.

Bar-and-symbol Chart. This is merely a simple bar chart with supplementary information indicated by a cross-line, circle, diamond, or some other symbol.

Subdivided-bar Chart. This type of bar chart, like, the 100% bar chart described in the following statement, is also referred to as a ''segmented-bar'' or ''component-bar'' chart. The scale values of the subdivided-bar chart are shown in absolute numbers. To portray percentage distribution of components, the 100% bar chart should be used.

Subdivided 100% Bar Chart. This type of chart consists of one or more segmented bars where each bar totals 100%. The various divisions of the bars represent percentages of the whole.

The *strength* of charts and graphs is their ability to clarify, emphasize, or confirm the significance of a set of facts or statistics. The *weakness* of charts and graphs is their potential for misuse. Through carelessness or deliberate distortion graphs/charts can misrepresent or confuse relationships of data.

Each type of graph has its own advantages and disadvantages. Most graphs/charts are variations on one of two types: the *bar graph* and the *line graph*.

Bar Graphs/Charts

According to Calvin and Stanton Schmid, authors of the *Handbook of Graphic Presentation* (New York: John Wiley and Sons, 1979), the *bar chart* is especially useful in comparing the magnitude or size of "coordinate items or parts of a total." The length of each bar "determines the magnitude or value of a series of categories." *Vertical bar charts,* or *column charts,* are best for depicting amounts in a time series and contrasts in amounts for two or more time series.

The eight common types of bar charts and the eight common types of column charts appear with brief explanations of each in Figures 6.6 and 6.7 respectively.

Figure 6.6   (*Continued*)

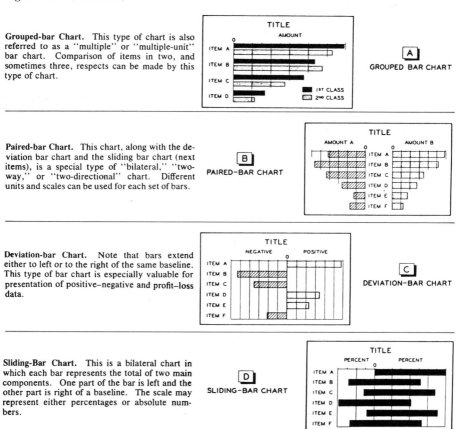

Grouped-bar Chart. This type of chart is also referred to as a "multiple" or "multiple-unit" bar chart. Comparison of items in two, and sometimes three, respects can be made by this type of chart.

Paired-bar Chart. This chart, along with the deviation bar chart and the sliding bar chart (next items), is a special type of "bilateral," "two-way," or "two-directional" chart. Different units and scales can be used for each set of bars.

Deviation-bar Chart.   Note that bars extend either to left or to the right of the same baseline. This type of bar chart is especially valuable for presentation of positive–negative and profit–loss data.

Sliding-Bar Chart. This is a bilateral chart in which each bar represents the total of two main components. One part of the bar is left and the other part is right of a baseline. The scale may represent either percentages or absolute numbers.

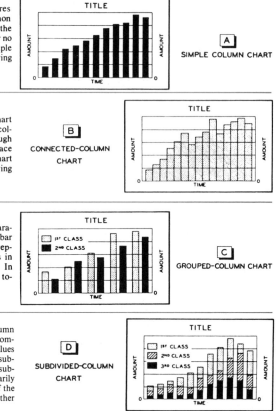

Simple Column Chart. In several basic features the simple column chart has much in common with the simple bar chart. The baseline of the column chart is drawn horizontally, and under no circumstances should it be omitted. The simple column chart is particularly valuable for showing time series.

Connected-column Chart. This type of chart possesses characteristics of both the simple column chart and staircase surface chart. Although all the columns are distinct, there is no space between them. The connected-column chart may be particularly valuable as a space-saving device.

Grouped-column Chart. This chart is comparable to the grouped-, multiple-, or compound-bar chart. Two or occasionally three columns representing different series or different classes in the same series can be grouped together. In grouping the columns they may be joined together or separated by a narrow space.

Subdivided–column Chart. The subdivided–column chart, like the subdivided-, segmented-, or component-bar chart, is used to show a series of values with respect to their component parts. The subdivided-column chart is also similar to the subdivided-surface chart. Crosshatching is ordinarily used to differentiate the various subdivisions of the columns. The scale may be expressed in terms either of absolute or relative values.

Figure 6.7   Eight types of column charts. (From Calvin F. and Stanton E. Schmid, 1977, pp. 80–81.)

Line Graphs/Charts

On a line graph the plotted points of the data are connected by a solid or broken line. The movement of this line shows the variations in a trend. Mary Eleanor Spear, a graphic analyst formerly with the Bureau of Labor Statistics, advises when to use which kind of line graph in Figure 6.8.

Different kinds of bar, column, and line graphs can and should be used separately or in conjunction to stress different aspects of the data presented in tables or to demonstrate different implications of the same aspect. In choosing which graph(s) to use, you need to know: (1) which perspectives or implications of the data you wish to emphasize, and (2) which type of graph best represents what you want to show. In Figure 6.10 Schmid and Schmid give eight graphic representations of all or part of the information provided by table in Figure 6.9. Each graph represents a slightly different approach to, or places a slightly different emphasis on the same data. Line graph B (Figure 6.10), for example, demonstrates patterns over a six-month period of the basic fact verified by the

Net-deviation-column Chart. The net-deviation- and the gross-deviation-column charts are similar to the bilateral-bar charts. They emphasize positive and negative numbers, increases and decreases, and gains and losses. In the net-deviation chart the column extends either above or below the referent line, but not in both directions.

Gross-deviation-column Chart. The columns in this type of chart extend in both directions from the referent line. By means of crosshatching, both gross and net changes can be readily portrayed.

Floating-column Chart. The floating-column chart is a deviational or bilateral chart with 100% component columns. The deviations from the referent line represent positive and negative values or differential attributes.

Range Chart. The range chart shows maximal and minimal values in time series. This chart has been referred to as a "stock-price" chart, since it is extensively used in plotting highest and lowest daily, weekly, or monthly stock quotations. Average values also can be readily indicated on the columns.

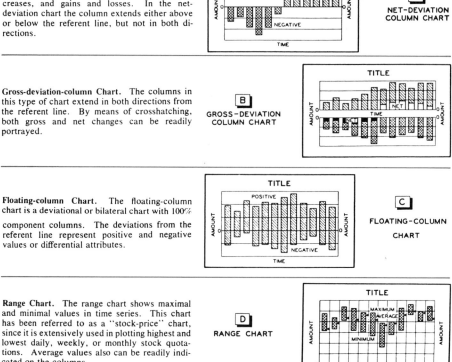

Figure 6.7 *(Continued)*

table in Figure 6.10—that most of the time, lack of parts is a more frequent cause of vehicles being disabled for four days or more than any other cause considered. Bar graph G, in contrast, expresses the percentage of each kind of vehicle disabled four days or more due to lack of parts. It facilitates such observations as the fact that more medical service than quartermaster vehicles are out of service beyond the minimum length of time due to the parts problem.

Guidelines for Preparing Graphs/Charts

- *Plan ahead for visual clarity.* There should be sufficient space, for instance, between bars on a bar graph. All bars should be the same width, and the space between each should be roughly half the width of the bar.
- *Label and order parts logically* by alphabetical, numerical, chronological, qualitative, geographic, or progressive sequence.
- *Choose the kind of line, bar, or column chart* that most accurately represents the aspect of the data that you want to demonstrate.

- *Avoid contracting; expanding; or skipping the original grid.* For example, in Figure 6.11, chart A demonstrates what happens when an irregular time sequence (chart B) is "corrected." The upward trend in chart A, where the grid size for each year has been made uniform, looks much more dramatic than it actually is.

- *Start with a zero-base line to avoid distortion.* For example, in Figure 6.12, charts A and B illustrate the same information on average weekly earnings, but look very different because the baseline of A begins at the lowest average weekly earning ($60), while the baseline of B begins at zero.

- *Use drawing techniques that will not distort the plottings on your graph.* Avoid elimination of grid marks or crosshatching. In Figure 6.13a the plottings on lines A and B are identical for 1955 and 1960, but without the grid, they seem different. And in Figure 6.13, we see the confusing effects of crosshatching.

Diagrams and Illustrations

The purpose of diagrams and illustrations is to clarify explanations, instructions, or designs by allowing readers to see what they are reading about. An effective diagram or illustration provides just enough detail and labeling to suit the readers' expertise and the purpose for which the illustration will be used. At the same time, it does not provide so much detail that the viewers are overwhelmed.

The *strength* of a diagram or illustration is that it can facilitate reader comprehension of the text it accompanies; the *weakness* is that a poorly drawn diagram, a blurry photograph, an inadequately labeled drawing, or an illustration that is inconsistent with the accompanying text can do more damage than good to readers' understanding.

Guidelines

- *Make sure that diagrams print clearly, that labels are legible and properly sized, that the information on the diagram is consistent with the information given in the text or tables.* The companion table and diagram in Figure 6.14 are a good example of how graphics can enhance comprehension. Together, table and diagram make the written explanation of types of lines recommended by the British Standards Institution easier to understand.

- *Make sure that diagrams illustrating "before" and "after," or possible variations on equipment, represent changes clearly and precisely. Label added or altered parts simply and briefly.* For example, by examining Figure 6.15 even readers not familiar with solenoids can grasp what is involved in converting a pull-type into a push-type. (Solenoids are electric switch assemblies consisting of a cylindrical coil of insulated wire and a metal core free to slide along the coil axis under the influence of the magnetic field.)

Figure 6.8 When to use line charts. (From Mary Eleanor Spear, *Practical Charting Techniques*, New York: McGraw-Hill, 1969, pp. 74–75.)

WHEN TO USE LINE CHARTS

1. *When data cover a long period of time* (A)
2. *When several series are compared on the same chart* (B)
3. *When the emphasis is on the movement rather than on the actual amount* (C)
4. *When trends of frequency distribution are presented* (D)
5. *When a multiple-amount scale is used* (E)
6. *When estimates, forecasts, interpolation, or extrapolation are to be shown* (F)

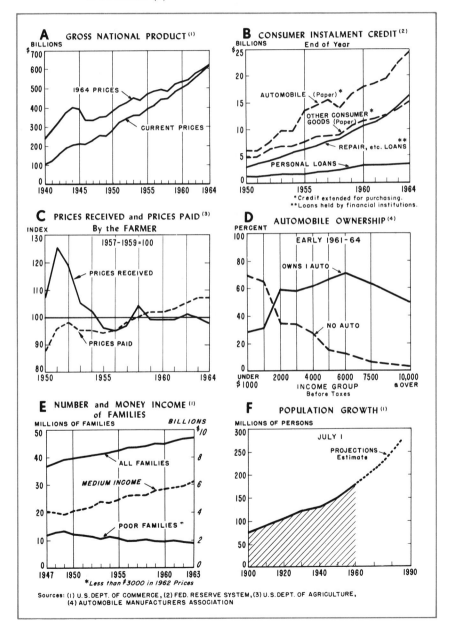

Sources: (1) U.S. DEPT. OF COMMERCE, (2) FED. RESERVE SYSTEM, (3) U.S. DEPT. OF AGRICULTURE, (4) AUTOMOBILE MANUFACTURERS ASSOCIATION

Table 1-2 General-Purpose Vehicles Disabled 4 Days or More

Service	May		June		July		August		September		October	
	Total	Due to Lack of Parts	Total	Due to Lack of Parts	Total	Due to Lack of Parts	Total	Due to Lack of Parts	Total	Due to Lack of Parts	Total	Due to Lack of Parts
All services	922	625	1271	774	856	533	981	675	1247	679	1486	682
Chemical	98	48	78	43	78	23	39	17	91	51	95	51
Engineers	136	116	136	97	29	29	19	18	29	29	21	17
Medical	41	25	81	44	4	2	6	5	5	2	9	6
Ordnance	252	174	368	221	265	193	353	268	606	218	654	164
Quartermaster	116	63	76	41	61	21	61	23	59	31	58	28
Signal	69	44	80	34	28	15	31	23	39	34	46	30
Transporation	210	155	452	294	391	250	472	321	418	314	603	386

Figure 6.9 Table supplying the data for the graphs in Figure 6.10. (From Schmid, 1977, p. 80.)

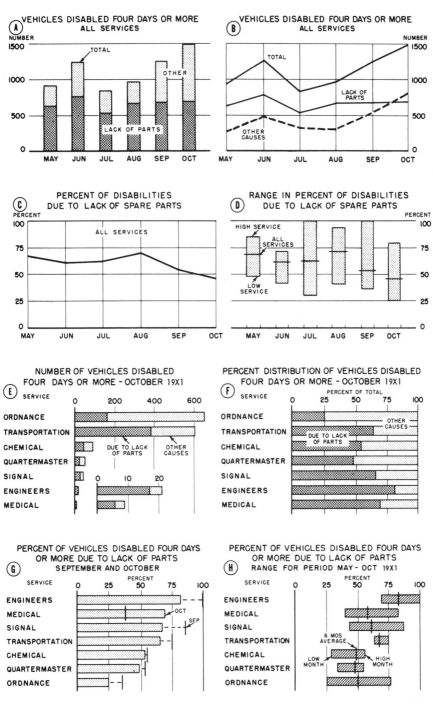

Figure 6.10 Eight graphs using the information in Figure 6.9. (From Schmid, 1977, p. 80.)

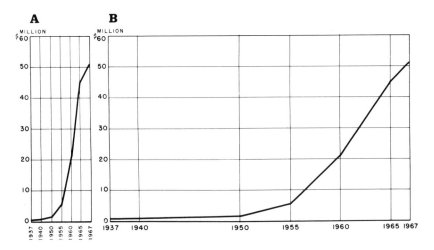

Figure 6.11   Spacing an irregular time sequence. (From Spear, 1969, p. 57.)

- *Represent even very complex concepts by simple illustrations.* For example, the problem of demonstrating relationships between brief time spans (such as several milliseconds) is well-handled in Figure 6.16 which shows how advance notice of a direct-current (dc) power loss allows orderly power shutdown in a data processing system.
- *Label all parts thoroughly and clearly.* For example, the illustrated instructions in Figure 6.17 describing how to attach a grass bag to a lawn mower are more difficult than they need be. The reader's eye must flit from the instructions to the parts list to the diagram, to understand fully the significance of the labeled parts. (Also, can you find the "channel" referred to in the instructions on either the diagrams or the parts list?)

Figure 6.12   The broken amount scale. (From Spear, 1969, p. 59.)

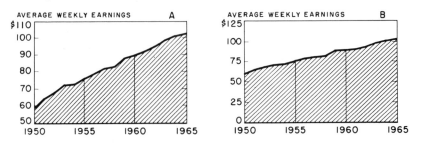

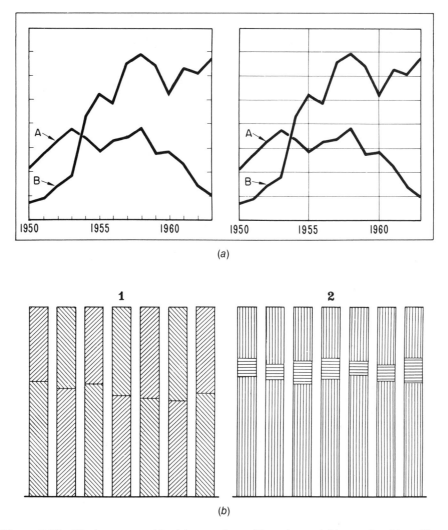

Figure 6.13 Illusions created by (a) removing grid marks, and (b) crosshatching. (From Spear, 1969, pp. 60–61.)

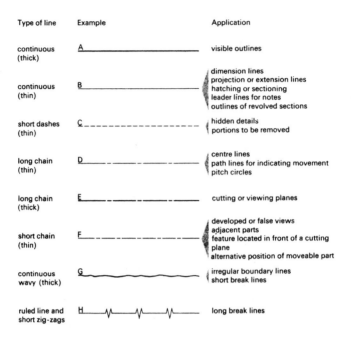

Type of line	Example	Application
continuous (thick)	A _____	visible outlines
continuous (thin)	B _____	dimension lines projection or extension lines hatching or sectioning leader lines for notes outlines of revolved sections
short dashes (thin)	C - - - - - - - - - -	hidden details portions to be removed
long chain (thin)	D ___ _ ___ _ ___	centre lines path lines for indicating movement pitch circles
long chain (thick)	E ___ __ ___ __ __	cutting or viewing planes
short chain (thin)	F _ _ _ __ _ _ _ __	developed or false views adjacent parts feature located in front of a cutting plane alternative position of moveable part
continuous wavy (thick)	G ∿∿∿∿∿	irregular boundary lines short break lines
ruled line and short zig-zags	H ─/\─/\─/\─	long break lines

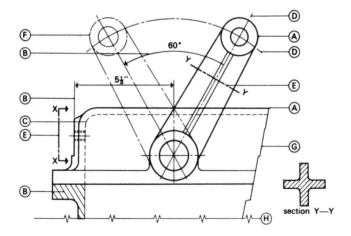

Figure 6.14 Types of lines recommended by British Standards Institution for engineering drawings. (From Arthur Lockwood. *Diagrams: A Visual Survey*, New York: Watson-Guptill, 1969.)

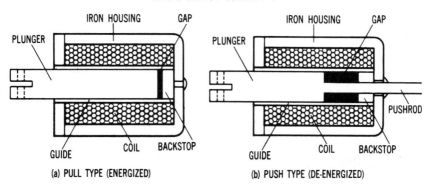

TYPICAL SOLENOID CONSTRUCTION

By attaching a pushrod to the plunger through a hole in the backstop, the pull-type solenoid (a) becomes the push type (b).

Figure 6.15 When a pushrod is attached to the plunger through a hole in the backstop, the pull-type solenoid (a) becomes the push-type (b). (From David Luckenbach, "Solenoids," *Electronic Products,* 23, April, 1981, p. 67.)

Diagrams and Photographs for Technical Manuals

In an article entitled "Developing Illustrations for Technical Manuals from a Writer's Viewpoint" (*Technical Communication,* 26 [2nd Quarter, 1979], pp. 4–9) Ralph S. Miller and Edward S. Read, Jr. provide a table of typical manual sections (Figure 6.18) and a discussion of how to decide on and locate the kinds of illustrations necessary for each. Schematics; wiring diagrams; logic, block, and timing diagrams; mechanical drawings; parts-locating views; and waveforms each serve a special purpose. For example, if a technical manual covers a single unit or module, a photograph will work for the first illustration or frontispiece. But for spread-out, multi-unit systems that cannot be photographed effectively, the reader requires a perspective drawing (Figure 6.19).

The final illustrations that appear in manuals are often developed by the editor/writer/illustrator from "raw input," rough sketches by the engineers who designed or built the equipment. Figure 6.20 shows a sample rough sketch and the clear illustration developed from it. The refined version improves the clarity of the graphic by adding in all modules, cabinets, wiring, and cables related to the parts directly relevant to the signal path, and by making the distinctions between different parts more readily visible to potential troubleshooters.

As you can see, there are many variations on the basic tables, graphs, diagrams and pictures; and which ones you use depend on the effect you want to achieve, or the point you want to stress or illustrate, and on what facilities you have. If you intend to be a technical writer—or, if for any reason you need to create, find, or adapt illustrations, take one or more courses in drafting, mechanical

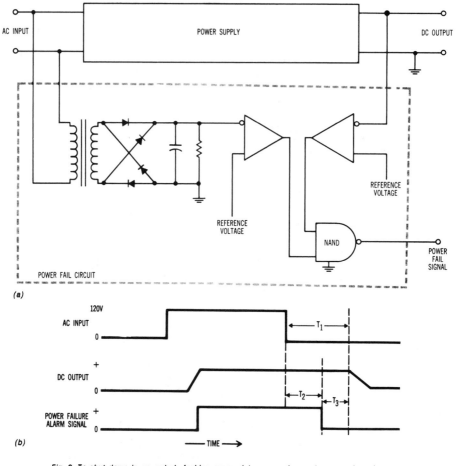

Fig. 2. *To shut down in an orderly fashion, many data processing systems require advance notice of a dc power loss. Typical power fail circuits (a) monitor both the input and output and, at turn on, produce a logic 0. When the supply's output is in regulation, a logic 1 is produced (b). Since most supplies can deliver power for several milliseconds (T_1) after loss of input power, a delay (T_2) is introduced between the loss of the ac input and the transition back to logic 0. Even so, there is enough time (T_3) for the load to accomplish an orderly shutdown.*

Figure 6.16 Relationships between brief time spans. (From Walter J. Hirschberg, "Power Supply Bells and Whistles," *Electronic Products*, 23, April, 1981, p. 48.)

drawing, engineering graphics, and/or computer graphics. At the very least, read several articles, journals, and books on graphic design, such as:

Graphic Communications World, A. E. Gardner, ed., vol. 12, nos. 7–12, April 3–June 18, 1979.

Hanna, J. S., "Six Starts Toward Better Charts," *Technical Communication*, vol. 29, #3 (3rd Quarter, 1982), 4–8.

SKIL   MODEL   518

LAWN MOWER
GRASS BAG ATTACHMENT
NO. 302325
ROTARY MOWER

Assembly Instructions

Fig. 1 — Join upper support arm "A" to lower arm "B" by slipping rod ends into tubes. Place frame "C" under bar welded to lower arm and bolt with (2) 10–24 × 1/2" screws and nuts. Insert screws from bottom side as shown. NOTE: Frame can be put on backward. Check it against the opening before assembly to arm.

Fig. 2 and 3 — Upturned rod ends slip into deck bracket as shown. Frame seats on top of discharge opening and butts against the toe guard.

Stretch elastic bag mouth over top of arm assembly and pull down to channel frame. Elastic mouth fits into the channel.

Parts List

When ordering parts always list the part number and the name of the part.

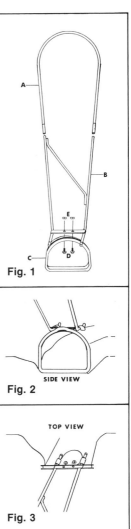

Fig. 1

SIDE VIEW

Fig. 2

TOP VIEW

Fig. 3

Skil Part No.	Item No.	Description	Number Required
73068	A	Upper Arm	1
302322	B	Lower Arm	1
302323	C	Flange Frame	1
69085	D	10–24 × 1/2" Bolts	2
67682	E	10–24 Keps Nuts	2
302324		Bag	1
302326		Deck Bracket (packed separately)	1

Figure 6.17   How to attach a grass bag to a lawn mower.

Lockwood, Arthur, *Diagrams: A Visual Survey of Graphs, Maps, Charts, and Diagrams for the Graphic Designer,* New York: Watson-Guptill, 1969.

Lutz, R. R., *Graphic Presentation Simplified,* New York: Funk & Wagnalls Company, 1949.

Plan and Print, vol. N55, no. 1 (Jan., 1982), special issue on computer-aided design (CAD).

Table 1—Transmitter Manual Outline

Section	Title	Purpose
1	GENERAL INFORMATION	What it looks like
2	OPERATION	How to operate it
3	PRINCIPLES OF OPERATION	How it operates
4	SCHEDULED MAINTENANCE	How to keep it operating
5	TROUBLESHOOTING	How to diagnose faults when they occur
6	CORRECTIVE MAINTENANCE	How to fix faults
7	PARTS LIST	How to locate parts for repair
8	INSTALLATION	How to site and make initial installation

Figure 6.18   Table of typical manual sections. (From Ralph S. Miller and Edward S. Read Jr., "Developing Illustrations for Technical Manuals from a Writer's Point of View," *Technical Communication*, 26, 1979, p. 4.)

Schmid, Calvin F., and Schmid, Stanton E., *Handbook of Graphic Presentation*, 2nd ed., New York: John Wiley and Sons, Inc., 1979.

Spear, Mary Eleanor, *Practical Charting Techniques,* New York: McGraw-Hill, 1969.

Write, J. V., "Draftsmanship," *Folio,* vol. 10, no. 11 (Nov., 1981), 112–117.

Figure 6.19   A perspective drawing. (From Miller and Read Jr., 1979, p. 6.)

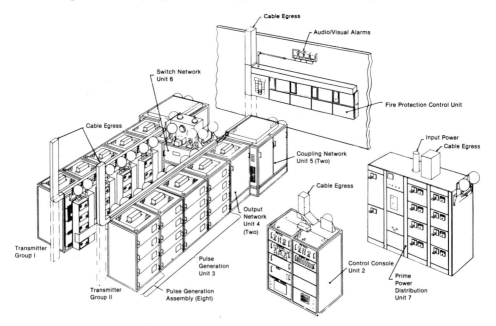

Frontispiece of Loran-C transmitter manual.

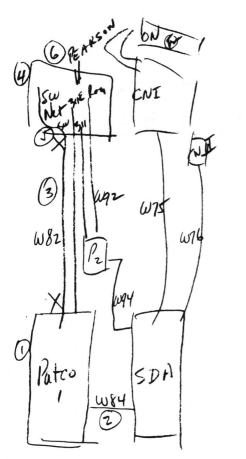
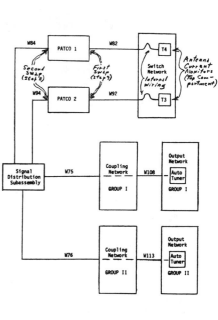

Figure 6.20 Rough sketch versus finished drawing. (From Miller/Read Jr., p. 7.)

And remember, whenever you read a textbook, professional journal, report, or manual, notice which graphic techniques are most effective, and ask yourself why.

EXERCISES

1. Select a technical journal in your field or one from a list provided by your instructor. Go through two or more articles that use graphics to analyze (a) the effectiveness of the illustrations used, and (b) whether or not illustrations that would have been useful are lacking.

2. Select one of the unillustrated sample reports in this text or one of your own. Decide where illustrations could be used to make it more effective. What kind of graphics would you use and why?

The Writing Process: After You Write

The revising process takes place before and after reader feedback. Before your communication is distributed, re-read it twice—once to make sure the strategy, format and language are adapted to the audience, and a second time to check for stylistic and grammatical errors such as confusing word order, ambiguous meaning, distracting repetition, misspelling or misuse of words, or lack of parallel construction in lists.

Revise in response to reader feedback by either making requested changes based on new information, or explaining why the changes should not be made. Always get requests for changes in writing or write them out yourself in a memo to the requesting party. If you disagree with the request, put your explanation in writing and send copies to project participants.

REVISING BEFORE DISTRIBUTION

When you have completed writing what you believe to be the final draft of a technical communication, read it through twice. The first time, try to make sure that your writing caters consistently to your audience's perspective. Have you used language that is neither condescending nor incomprehensible? Have you included visual aids where your audience requires them? Have you left out explanations that readers will need, or included lengthy descriptions that they could do without? After correcting for reader adaptation, read through a second time to correct style and grammar. Both readings and revisions may be easier for you if you compose on a computer using a word processing program. Instead of cutting and pasting, you just push the right buttons to move, replace, or correct one letter or whole sections of your text.

Revising for Suitability to the Reader

Figure 7.1 is a draft of a proposal written by a Research and Development engineer in a technical writing seminar. His audience is the Board of Directors, who

THE "UP-SIDE DOWN LIGHT"

The proposed "upside-down light" is a new-generation solution to the age old problem of providing high quality lighting of open spaces while conserving energy. Unlike conventional lighting products, it provides high visibility at the desk top while maintaining visual comfort for the user. Upon maturation of this product line, the division profits on lighting will have increased by 108 percent on increased lighting sales of 75 percent. This new product will replace the existing products, resulting in improved plant performance and overcoming the customer objections to our aesthetically undesirable products. All this is provided with a one-time investment for development of $375,000 at an ROI (return on investment) of 37.5 percent.

Upside-down lighting is a new concept created within our R & D department; patents have been filed on the concept and the design. The concept of reversing the light came in response to customer dissatisfaction with conventional lighting where direct light causes glare, discomfort, poor visibility, and reduced worker performance while using a disproportionate amount of electrical energy. Prototype testing shows the upside-down light to provide reduced glare or discomfort, increased illumination and 30 percent less energy requirements.

Although the concept is unique, the basic mechanical design will allow manufacture by our existing techniques. Thus, the new design will cause minimal disruption of our manufacturing facilities and short changeover times.

New designs usually result in increased costs, especially in development, which is true in this situation also. But the increased costs can be absorbed in increased selling price because of the apparent value to the user. Based upon market surveys with key accounts, it is estimated that selling price per unit can increase $12.50 (28%) while manufacturing cost increases $3.10 (9%).

Market surveys further show that unit sales will increase by 60 percent in the first full year of sales due to satisfying of customer needs, increased market activity, and the addition of new market segments. The combined sales price increase and unit increase results in a volume increase of $1.2 million (75%) giving a net pre-tax profit increase of $300,000 (108%).

Figure 7.1 Proposal draft.

will have the final say on whether the new "upside-down light" will replace some of the company's existing products.

On re-reading his report from the intended audience's stance, the first thing the writer notes is the language and tone of the opening paragraph. Yes, all writing is selling, but is the style of ad copy appropriate for the Board of Directors? Describing the upside-down light as a "new generation solution" is meaningless and possibly even offensive language to the Board of Directors, who want to know in a nutshell:

- What does it do?
- What does it cost?
- How much of a profit can we make?

It may be that the writer's first sentence was the "false start" that he needed to get into his subject. But it is not the start that his audience needs: he will have to revise it. Having decided that the opening sentence needs changing, the writer looks at the rest of the opening. Does it summarize all the main points that the writer makes about the proposed lamp in the remainder of the report? Does the order of presentation of the opening summary correspond to the order of presentation in the body?

To answer these questions, the writer tries to interpret his writing as his readers would. The opening gives the readers six reasons why they should add the upside-down light to their product lines, in this order:

- It conserves energy;
- It provides high visibility without causing visual discomfort;
- It will increase division profits eventually by 108% and light sales by 75%;
- It will improve plant performance by replacing some existing product;
- It will overcome customer objections to aesthetically undesirable products;
- It requires a one-time investment of $375,000 at an ROI of 37.5% and an ROICU of 18%.

In the body of the report four reasons to manufacture the upside-down light are presented in the following order:

- It provides reduced glare and lessens reader discomfort;
- It can be produced using existing equipment and facilities with the minimal disruption and changeover times;
- It will cost a lot to develop, but because of its consumer desirability, it can be marketed at a price that will recover these additional costs;
- In the first full year of sales, it will increase unit sales by 60% and result in a pre-tax profit increase of $300,000 (108%).

Placing these lists side by side, the writer can see that the opening and body of his report look like two different approaches to the same material rather than a summary followed by a more detailed version of the same information. The writer can see that the answers to the main questions his readers will ask are in the report, but need to be rearranged. He finds that by rewriting the opening as a synopsis of the existing body, he can make the entire report a more effective communication:

The proposed upside-down light designed by Research and Development solves the problem of customer dissatisfaction with similar products that cannot provide adequate direct light without painful glare. The prototype provides increased illumination along with decreased energy requirements. It can be manufactured with existing structures, and although it will require a one-time investment of $375,000 for development, it will compensate with a projected ROI of 37.5%, ROICU of 18%, and increase unit sales 60% the first full year.

Headings

Following this revised opening, the writer keeps the original body with one significant change. He adds headings to help the reader see where the detailed discussion of each of his four main points begins and ends:

1. *A solution to poor visibility of present high intensity lamps.*
2. *A new product that will increase manufacturing efficiency.*
3. *A new design that will quickly recover its development costs.*
4. *An attractive product that will increase unit sales and profit per year significantly.*

The final thing the writer notes in assuming the readers' point of view is that he has left them hanging by omitting a conclusion. He adds a conclusion by restating the points made in his four headings, and his revision for reader comprehension is completed.

Revising for Style and Grammar

You will want to do a second separate reading of your report for another purpose—stylistic and grammatical changes. Often, on a second reading, you find you can shorten sentences, add commas, and catch other communication distractors that you did not notice the first time through.

It may seem unreasonable to you to read through what you have written twice, especially when your report is long. But you cannot concentrate fully on whether you have properly adapted your material to your audience if you are dotting "i's" and correcting misspellings at the same time. As you will see when we discuss revising in response to reader feedback, avoiding communication problems in business *before* they occur is much easier and much less time consuming than trying to fix up misunderstandings and compensate for disappointments after the fact. What you are aiming for is not the achievement of grammatical perfection, but the avoidance of those kinds of misusage which call attention to themselves and away from the point they are meant to express.

For example, can you identify the "communication distractions" in the following excerpts?

Example 1: Zinc Concentrate Study

SUMMARY

The zinc content in the mined ore has been falling for 10 years. This falling grade has prompted numerous questions on the best alternative to maximize profits from the zinc in the ore. This is a complex question that requires detailed evaluation. This short memo is designed to develop the parmaters for a through review of zinc concentrate production. The report will be completed by June, 1981.

In this passage there are at least three distractions that, if eliminated, would improve effective communication.

1. *Number inconsistency.* In sentence two "numerous questions" have been raised. In the next sentence, the problem is referred to as a "complex question." The reader does not know if the question of how to maximize zinc profits is one of several questions or itself consists of several subsidiary questions.

2. *Spelling.* In sentence four, two words are misspelled: "parmaters" and "through." The misspelling of "parameters" in particular is distracting because it is a key concept in the report, and the reader may not be sure if the word is what is seems to be.

3. *Word repetition.* Of the five sentences in the opening statement, three begin with the word "this." Patterned repetition of minor, vague words pulls attention toward them and away from your key points.

Example 2: Recommended Repairs on Acid Tank

a. Replace tank bottom with ⅜" thick mild steel plate
b. Porosity holes in ring weldments should be cleaned and rewelded.
c. Discharge handle and plug mechanism
d. According to company specifications, tank repainting.

This passage contains two major errors that interfere with communication.

1. *Parallel construction.* The key interference here is failure to employ parallel construction in list making. If you started the first recommended repair as a command, use the command form consistently, or use passive construction throughout (for example: "tank bottom should be repaired"). Whatever phrasing pattern you use, repeat it for all instructions listed. The repetition of word order in this case is not distracting. Instead, it facilitates understanding of the main points and of how they are connected. Parallels in grammatical structure reinforce relationships in content. For example, the repairs list might be revised as follows:

Recommended Repairs
1. Replace tank bottom with ⅜" thick mild steel plate.
2. Clean and reweld the porosity holes in the ring weldments.
3. Repair the discharge handle and plug mechanism.
4. Repaint the tank according to the company's specifications.

(Notice that each item in the list is now end-punctuated consistently and correctly—with a period.)

2. *Ambiguity.* In this example, changes in word order also result in ambiguity. For instance, in item 3 in the unrevised version, it is unclear what should be

done to the discharge handle and plug mechanism. It also may be unclear whether "discharge" is being used as an adjective to describe the handle or as a verb to state what should be done to the handle and plug mechanism.

Example 3: Preventive Maintenance Program

A preventive maintenance program is proposed to be set up to reduce costs and to assure continued operation of the Surface Mine Department equipment. This program would cover three (3) types of equipment: hoists, fans, and compressors. This program would be performed weekly and scheduled with the general mine foremen according to mine production demand. This program would cost $50,000 in materials and downtime.

This passage exhibits two major problems.

1. *Awkward word order.* In the first sentence of the proposal, the chosen word order necessitates a distracting repetition of "to," and turns a simple point into a complex statement. Revised, it might read: "A preventive maintenance program is proposed that would reduce costs and ensure continuous operation of the Surface Mine Department equipment (SMDE).

2. *Distracting repetition.* Three of the four sentences in the opening statement begin with "This program," giving it a distracting "Dick and Jane" format. If you have two or more points to make about the same program, process, or object, combine them in one sentence and use numbers to distinguish them. The four sentences of this opening, for example, could be combined into one:

A Preventive Maintenance Program (PMP) is proposed that would

1. ensure continued operation of the Surface Mine Department's hoists, fans, and compressors,
2. reduce equipment costs,
3. be performed weekly and scheduled with the general foreman according to mine production demand, and
4. cost $50,000 in materials and downtime.

In the revised version, unnecessary and distracting words are omitted, and the main points are featured. Numbering and parallel construction help the reader organize and master the pertinent facts quickly.

Example 4: How to Play Golf

When you step up to hit the ball, this initial movement is called the address.

Whether you are right-handed or left-handed depends on what hand overlaps with which finger.

Your next position of concern is stance comfort. By that I mean you have to adjust the distance as to how far you want to stand away from the ball, to whatever is comfortable for you.

Describing and instructing are major functions of all technical writers. This student describing his favorite sport fell into the typical traps of describers—wordiness and ambiguity. The results are more amusing than informative.

The first sentence can be made shorter and to the point:

Stepping up to hit the ball is "the address."

The second sentence seems more complicated than it is. By substituting one correct word for one misused word, the writer makes his meaning clear:

Whether you are right-handed or left-handed *determines* what hand overlaps which finger.

The last sentence needs to be rephrased to avoid unnecessary repetition. If you tire your readers out comprehending your simpler points, they will not have any energy left for your really complex messages. Some possible rephrasings are:

Stand only as far away from the ball as is comfortable for you,

or

How far you stand from the ball is determined by what is comfortable for you,

or even more concise:

Players' distance from the ball depends on individual comfort.

COMMON STYLISTIC AND GRAMMATICAL ERRORS

As you can see from the four samples above, the most common communication interferences that you will need to recognize and revise are:

1. Confusing Word Order
2. Ambiguous Meaning
3. Distracting Repetition
4. Misspelling or Misuse of Words
5. Lack of Parallel Construction in Lists.

If, on the second reading of your communication, you use these five criteria alone, you can catch and revise the major communication distractions in your reports, letters, or memos before you pass them on to your readers. Practice on the exercises at the end of this chapter and on the written communications that are sent to you.

REVISING IN RESPONSE TO READER FEEDBACK

When you hand in a paper for a class, you are more or less done with your part in it. You will get it back with a grade, corrections, and comments, and you may even discuss the comments with the instructor. Occasionally, you will be allowed or required to re-write the assignment. More often than not, however, in the classroom situation there is little opportunity to make practical use of your readers' response.

In the business world your situation is markedly different. Projects are taken up, slowed down, revised, or dropped based on what you and others write. People are hired or fired, divisions are created or disbanded, companies are purchased or passed by—based on written input from you and your associates. If your part of the writing task keeps things smoothly in motion, you get a pat on the back and a favorable notation in your service record. If your writing or others' interpretations of your writing interfere with getting things done, however, you are going to hear about it, and you are going to have to respond quickly and acceptably—often, by revising what you have written in response to general or specific requests for changes by one or more of your intended (or, perhaps, unanticipated) readers.

You are more likely to respond in ways that are least compromising for you and most acceptable to your readers if, in each instance, you understand what the problem is and what your options are. Most of the possible reader responses come under one of the two situations discussed below. As a general rule, however, regardless of what situation you are in: *Get the reader to put his/her objections or additional requirements in writing.*

If the response reaches you over the phone or verbally, in the hallway or restroom, write up your version of what was said in the form of a memo addressed to the requesting party. Make sure that all the principals concerned receive copies.

The Two Most Frequent Reader Feedback Situations and How to Respond

Situation 1: Integrating New Information

What Happens: You are asked to rework your report taking into consideration aspects of your readers' perspective that you had not considered in your initial planning, or providing more or different information.

What to do: Use this information about your readers' expectations to make your report more intelligible to them.

Example 1:
In a technical writing seminar conducted at a telecommunications company, students were asked to write a proposal. The intermediate audience was the writer's supervisor who would pass it on to decision-making upper-management personnel.

After the proposals were written, the students were divided into groups of six and asked to read four sample proposals as if they were the writer's immediate supervisor. The "supervisors" were to give the writer feedback on the suitability of the material for the intended audience.

One of the samples (Figure 7.2) was a proposal for early installation of the

Summary: Install IBM's newest version of TL3 during 3Q81 as part of an early-install program.

Discussion: We have tentatively decided to install TL3 soon after its availability in October 1981 for the following four reasons.

1. Performance enhancements: The enhancements appear to strongly favor a fairly large TSO shop, so we should obtain an improvement of 10–15% which will help us survive until the projected delivery of the 3Q81 in 3Q82.
2. TL3 is the first effort of system production to support the hardware enhancement due to be installed in October 1981. Again, this will help our survival until the 3Q81.
3. TL3 has associated with it the JES2 component that integrates NJE at no additional charge. This will allow us to interconnect with other client sites as requested.
4. TL3 will be the most current system and the focus of IBM's future efforts.

Of course, there are reasons not to convert to TL3:

1. Large rework of our JES2 model is required: However, future maintenance will be simplified if the JES EXITS are used, and the JES2 part can be tested separately.
2. TL3 is untested: However,
   (a) testing on site should show up most problems
   (b) feedback from other sites should show up the rest
   (c) due to our hardware configuration all of the new features will not be needed.
3. TL3 costs more: However, the incremental cost difference is $110 a month with a base of $1350 a month. In addition, the cost of the NJE package, which we had assumed we would obtain was $450 a month. Hence, there would be a net savings of $1,340 a month.

While TL3 is presently expected to be available in October 1981, IBM has indicated that they would be receptive to a request from the LABS for an early install of TL3. I would like to strongly recommend that the LABS request early install of TL3 for the following reasons:

1. performance help
2. NJE capability
3. separate software and hardware changes
4. stage "state of the art"
5. IBM expertise to assist in conversion.

Figure 7.2 Proposal draft.

IBM "TL3." Representing the writer's supervisor, one of the workshop groups responded to the proposal with the following memo:

RE: TL3 Early Install Proposal

I am reluctant to pass this on to J. R. until you have provided the answers he will need to put this through. You make a fine case for installation after 3Q81 and answer any possible objections he might have. But, he's not going to get approval for *early* install unless he knows more details.

For example, what additional costs are involved? what man-hours? down-time? Will the IBM staff assist gratis? Will they train our staff? What other ongoing tasks will have to be redistributed, redesigned, or rescheduled to enable early installation? Will there be negative effects? Can these be compensated for?

I think you have a good idea, and I'm for it. But the proposal I pass on to J. R. and that he gets through the budget people will have to answer these pertinent questions.

Please revise and resubmit.

F. Lewis

The writer would do well to profit from this added information on reader expectations in order to re-write his proposal. The five reasons for early installation mentioned at the end of the proposal should be its focus and should be explained in terms of costs and benefits rather than what technical maneuvers are involved. The result will be a report that convinces J. R., gives him the documentation he needs to convince his supervisors, and gets the writer what he wants, too.

Example 2:

Another proposal for a different company contained a passage which one group acting as the supervisor/reader considered unclear to the decision-making reader without the addition of more information. The proposal is for the implementation of an Environment to Support Systems Testing (ESST) and the passage in question declares that ESST will interface with two existing systems:

THE GOAL

The goal of ESST is to help test developers test more accurately, more completely, and more quickly.

THE PROPOSED SOLUTION

ESST will add five support services to two existing tools: DAT and ATG. ATG is responsible for creating minimally complete sets of tests. DAT is responsible for executing tests accurately and quickly. ESST will provide a basis for using these tools together efficiently.

Three services are to support the development of large, complex test plans: Management Services to keep track of large, interrelated sets of tests; Process Control Services, to allow fast and easy incorporation of new test sets into existing test plans; and On-line Storage Services, to provide an execution environment for tests to mirror these development services.

To the immediate supervisor the way in which ESST will interface with the Automatic Test Generator (ATG) and Driver for Automatic Testing (DAT) to improve both is clear because he is directly involved with their operation. But because he knows that ultimate approval will come from a reader who is less directly familiar with these two systems, he will ask the writer to revise the proposal by adding a fuller explanation of ATG and DAT.

If your reader feedback request is for more information or more detail, you have two choices. You can either add the information where appropriate, or respond with a memo explaining one of the following:

1. Why the information is not available.
2. Why you think that contrary to your reader's opinion, the additional information requested is not essential to the decision-making process or may even confuse the process by overcomplicating it.

Situation 2: Changing the Conclusion

What Happens: You get reader feedback requesting you to make a change in the views or recommendations expressed in your report.

What to do: Re-analyze the situation:

a. If the reader's reaction really does alter your thinking, revise the report to reflect your change in views graciously, and give credit where due.
b. If you still really feel that your evidence supports your conclusions, write a memo re-affirming your position (I understand-what-you-mean-*but* . . .).
c. If you feel your reader is wrong, unreasonable, biased, etc., you are confronting a political situation:
   You can:

• make the requested changes to please your superior
• submit both points of view to a neutral third party
• write a diplomatic but firm reply to the objecting reader, explaining why you will not make the changes (try to make yourself sound right without making him/her sound wrong).

Example 1:
Being asked to re-assess your conclusions usually occurs when your reader, prompted by the material provided in your report, recalls related information, input that was unavailable to you when you first wrote—which, when you know it, will alter your conclusions. For example, an international sales manager for a recording instruments company received a proposal from a sales representative in Mexico for the display of one of the company's new analog recorders at a local exhibition. The salesperson in Mexico made a reasonable recommendation based on the information available to him at the time. But by the time she received the proposal, the intended reader possessed additional information which would affect the writer's suggestions. She replied that because not enough

of the optional features would be available for display at the time of the exhibition, the recorder should probably not be exhibited yet, and that they should opt for private showings at selected companies instead. Since the reader responded not only with new information but with a compromise alternative, the writer has an opportunity to revise graciously:

Dear _____:

Thank you for your prompt consideration of my request.

As you know, I was unaware of the delay in preparation of available options on the Y2700. In view of this difficulty, we probably should not exhibit at the September exhibition after all.

Based on our experience with the biophysical monitor, your alternative approach seems like a good one. However, I would not entirely rule out the advantages in sheer numbers that an exhibition or convention provides. I am currently investigating what public events are scheduled in late fall/early winter in this area at which exhibiting our Y2700 might be appropriate. I'll let you know what I find out.

Meanwhile, please keep me posted on when the recorder with all options will be ready to go.

Sincerely,

Because plant operations, budget restrictions, and numbers and skills of personnel in large organizations are constantly changing—or because deliveries of materials are interrupted, workers go on strike, or corporate strategy is revised—there are frequently legitimate reasons for your reader to ask for revisions of your written work based on information you could not have had at the time you were writing, or on conditions that were changing while your final draft was being typed.

When a reader provides new information that does alter your thinking, see how ingenious you can be about incorporating it into your existing format. Always be gracious about legitimate requests, and always explain the reasons for the changes and what the new information is in a cover letter to the revised report.

Example 2:

If the new information or the reader's objections to your interpretation of the available information do not alter your thinking, you have two choices. Suppose that a quality control engineer is asked to evaluate the reasons for the numerous failures and high cost of maintaining the control instruments on a nitrogen furnace. After making the necessary tests, the engineer decides that the problems of inaccurate temperature control readouts and of electrical and mechanical failures have two causes: old-fashioned, overcomplicated, and worn-out equipment; and excessive operator error. Her recommended solution is to replace the existing unit with a new system of control instruments that needs less maintenance and is less susceptible to human error (Figure 7.3).

If the reader accepts the writer's interpretation of the problem, the solu-

On October 12, 1978, I was requested to investigate the numerous failures and high cost of maintenance of the control instruments for the Nitrogen furnace. Tests show that the problem is caused by outdated equipment and mishandling. The solution is to replace the present system with a more automatic unit requiring less handling and less repair.

Nitrogen furnace temperature control instruments have been accused of inaccuracy in both readout and process temperature, frequent failure of mechanical as well as electrical components, and complex operation methods. These events of failure are more noticeable as their frequency increases.

The events divide into equipment failures and operator errors. Wear due to age and the mechanical nature of the instruments cause equipment failures; operator errors are a by-product of inadequate training. Most of our control instruments are built to be operated and maintained by trained electricians and knowledge of the complete system is required to operate the units safely and properly. Thus, operator errors are most likely when a production person attempts to adjust the instruments and provide a "Quick Fix."

Our capabilities for maintaining the present instruments are diminishing; and replacement components are limited at best. Also, not all of our present equipment is still manufactured; yet most of it will have to be replaced in the next few years.

The cost of maintaining the control instruments for the year 1986 totaled $18,000. The cost for a one-to-one replacement scheme amounts to only $24,000, if we were to invest in a totally new system to replace the above. We could easily capitalize the new system, which costs anywhere from $20,000 to $35,000 (refer to Chart 1 for a comparative listing of the three proposed systems), and we can recover their cost in two years or less.

Most of the $18,000 maintenance cost goes toward labor time. The design of this proposed new system is modulized so we could troubleshoot and fault-isolate the defective equipment efficiently. Another benefit is the low down-time required to repair because we could repair by replacement.

The new system would be strictly electronic. Any unauthorized adjustment to the equipment would not be possible. Only group leaders would be able to access the instruments by either keys or code input.

We have tentatively planned for the switch over to the new system to occur toward the end of the year; allowing us to perform some on-line debugging. However, the final designing of the system will start immediately after the approval of the project. The project will require about one year to complete.

A new system of control instruments will enable us to better aid the operator in the operation of the equipment and the reduction of error. The system will also allow us more accurate control and readout of the process temperature, eliminating confusion for the quality control (QC) engineers. It will also reduce scraps by having its own decision-making capabilities based on parameter feed by the firing parametric data.

Figure 7.3 Failures in the nitrogen furnace.

tion—replacement—seems feasible. But one of the readers (the supervisor of nitrogen furnace personnel, for instance) may not "buy" the quality control engineer's explanation. The supervisor of nitrogen furnace personnel might reply with a memo recommending more standardized procedures for maintaining and operating the existing equipment. He might suggest that it would cost less to train personnel in the proper use of the equipment rather than purchase new equipment. Training, he might explain, would give the operator a career incentive and reduce the high turnover of employees in his section, which accounts in part for excessive error in equipment handling.

Another reader, perhaps the purchasing agent of the company, might respond to the section on cost of repair and replacement by stating that he can get all the spare parts needed for the old unit at a warehouse outlet whose inventory list he has just received. Therefore, it would not really be only slightly more expensive to completely replace the existing unit.

If these and other reader responses to her report did not change the writer's resolve that replacement is the best solution, she would need to respond to reader feedback in a counter memo answering each of the objections raised with reasons why they should not alter the decision that her report recommends. She should stick to the facts. For example, her answer to the recommendations for better training of personnel as a less expensive solution might be a comparative table of costs and a chart showing the length of time the trained person would remain an operator in the nitrogen furnace section.

Responses to objections should never be attacks on the readers. Whenever possible, the objector's idea should be given some credit followed by a factual explanation of why it won't work, or of how it is efficient but not as efficient as what your report suggests. Your aim is to make converts, not enemies.

By the same token, respond only to the fact-based responses of your readers. Ignore, at least, in your written replies, unsubstantiated personal remarks such as "What would she know about electrical equipment anyway!" Resist the temptation to fight back. In technical writing, calm reason substantiated by clear facts is your only effective weapon.

Example 3:

On those rare occasions when your technical communication happens to get caught up in a power play or a conflict of personalities which involves you only indirectly, the decision on whether and how to revise your report involves more than just your knowledge of the facts and your ability to write. Whether you revise or don't revise, you are going to make someone unhappy. Only you can decide what is most important to you—doing what you believe in, regardless of how it affects your career; going with the present power structure or betting on an up-and-coming authority; or trying to "ride the middle" through some sort of compromise.

Although your decisions on how to revise in response to reader feedback will not always depend on factors completely within your control, you will always be better off if you understand the ramifications of the situation in which you are operating. Being able to see the facts and circumstances from your readers' perspectives as well as your own is always an advantage, after, as well as before, your report is written and distributed.

EXERCISES

1. Revising Before Distribution

a. For Suitability to Readers

On the following pages are three papers written for technical writing classes: two versions of a proposal by a student to the athletic director of his university (Figures 7.4 and 7.5), and a report by a junior quality control engineer to his immediate supervisor with a simulated response from the supervisor (Figure 7.6).

First, decide which of the two versions of the proposal you think is better and why. Then, for all three papers, do the following:

- Analyze the communication situation.
- Decide whether or not, given the situation, the format and style is suitable to the readers.
- Revise where necessary to improve readability, and be prepared to explain your reasons in class discussion.

b. For Style

Re-read the two reports examined in exercise 1 to catch and correct the 5 most frequent errors in style and word use that distract readers from the writer's meaning. Check for and eliminate:

- confusing word order
- ambiguous meaning
- distracting repetition
- misspelled or misused words
- lack of parallel construction in lists.

2. Revising in Response to Reader Feedback

To practice responding in writing to other people's view of your written work, use a real or simulated situation such as the following:

a. If within the last year or two you have received written comments on a paper handed in to an instructor, write a list and summary of those comments. If the feedback was given in a conversation or conference rather than in writing, create a summary of what was discussed from memory. Then, write two versions of a reply to the instructor: one in which you agree to make the suggested

Athletic Director
CWRU
Dear Sir,

   In the past there has been considerable attention given to the concept of life sports education at CWRU. There have also been numerous requests that CWRU start a sailing program. The reasoning behind the requests was sound; sailing is a life sport and would if offered improve the quality of life and educational experience of students at the university. These requests were not granted because of a lack of funding. Sailing is an expensive sport. Aside from purchasing a fleet of dinghies, there are slip fees, maintenance costs, storage costs, transportation charges, and insurance.

   In the past decade there has been a dramatic interest and growth in a relatively new sport, boardsailing, which has been growing almost exponentially. It has become an accredited sport and was an Olympic event in the summer Olympics in 1984. During the last few years it has also become an intercollegiate sport. CWRU could become a top ranked contendor in this sport. Because of its relative infancy as an intercollegiate sport, we have the opportunity to develop a strong program and attract accomplished boardsailors.

   At present the only water sports offered at CWRU are swimming and one scuba class. The university is located only a few miles from Lake Erie which offers perfect conditions for Olympic-style boardsailing: good wind and relatively smooth water with indiscernible tides.

   The equipment required would be approximately a dozen sailboards and three dozen sails. (Each sailboard should be equipped with a fathead sail for light wind, a marginal sail for heavy winds, and a regatta sail for competition.) The equipment would cost about $13,000. A major yacht manufacturer has agreed to donate the money for the equipment. Replacement costs would be under $3,000 a year.

   There would be no slip fees; there are virtually no maintenance costs involved; and boardsailing is considered one of the safest sports in the world, so insurance (liability) cost would be extremely low. The only expense involved would be transportation; but the equipment can be moved easily in a station wagon or van.

   Boardsailing employs the basic principles of sailing only. A capable boardsailor is very well prepared to move up to larger crafts and quite often understands winds and hydrodynamics better than other classes of sailors.

   In conclusion, boardsailing offers a viable, inexpensive, and popular way of expanding the physical education curriculum, improving student life, attracting new students, and bringing favorable national, and possibly, international attention to CWRU and Cleveland.

Figure 7.4   Proposal for boardsailing team: version 1.

THE PROBLEM

Because our campus is located only a few miles from Lake Erie, and because sailing is a sport well suited to the concept of life sports education stressed at Case Western Reserve University, it has often been suggested that boating instruction and boating competition be part of our physical education program. Up until now, however, the prohibitive costs of purchasing and maintaining sailing equipment have prevented implementation of an otherwise highly desirable boating program.

THE SOLUTION

There is, however, a viable alternative which offers most of the advantages of sailboating and has none of the disadvantages. Boardsailing, a relatively new sport, has already become accredited and has been included in the summer Olympics since 1984. It employs all the principles of sailing, so that a capable board-sailor is well prepared to move up to larger crafts.

I propose that a course in boardsailing and an intercollegiate boardsailing team be established by the Physical Education Department in order to achieve the following benefits:

a. Boardsailing employs the basic principles of sailing, but is much safer and much less expensive.

b. Because boardsailing is a relatively new sport, a CWRU team would have an excellent chance to become a top-ranking contender, improve its public image, and attract athletically inclined new students.

Low Cost and Safety

The equipment required would be approximately a dozen sailboards and three dozen sails. (Each sailboard should be equipped with a fathead sail for light wind, a marginal sail for heavy winds, and a regatta sail for competition.) The equipment would cost about $13,000. A major yacht manufacturer has agreed to donate the money for the equipment. Replacement costs would be under $3,000 a year.

There would be no slip fees; there is virtually no maintenance cost involved; and boardsailing is considered one of the safest sports in the world, so insurance (liability) costs would be extremely low. The only expense involved would be transportation; but the equipment can be moved easily in a station wagon or van.

Increased Prestige and Enrollment

Boardsailing is in its infancy as an intercollegiate sport, and CWRU is in an excellent position to become a top contender. The university is located only a few miles from Lake Erie, which offers perfect conditions for Olympic-style boardsailing: good wind and relatively smooth water with indiscernible tides.

In conclusion, boardsailing offers a viable, inexpensive, and popular way of expanding the physical education curriculum, improving student life, attracting new students, and bringing national attention to our university.

Figure 7.5   Proposal for boardsailing team: version 2.

To: Tony Randall, Supervisor 1
From: Charles King, Junior QC Engineer
Re: Laser Cutting Tube Handler

The proposed four station horizontal tube handler appears to be the preferred method for high speed laser cutting of quartz tubing. Adaptability to the necessary techniques required for strain-free, dustless, smooth flat rolled edges is the most desirable feature of this horizontal method. Simplicity greatly reduces down time at size change-overs; new sizes require less time and fewer machine adjustments.

LIST OF ADVANTAGES TO HORIZONTAL METHOD

(1) Very simple machine at very low cost

(2) Has only a few unknowns

   (a) Loading

   (b) Parts detection

(3) No limitations to mechanical indexing, chuck problems

   (a) jaw grip

   (b) heat resistance

(4) Design is geared to the cutting needs

(5) Machine allows for maximum product mix

(6) Has great cutting capacity

(7) Minimum change-over time between sizes

(8) Would not require a highly skilled machine adjuster to keep it running

(9) Very quiet in operation

(10) Easy access to work on

The horizontal handler would solve many long-standing cutting problems that would be very costly to eliminate with the presently used vertical method. Cut quality has shown a marked improvement on samples run so far on a single chuck horizontal bench setup.

Versatility and simplicity of the horizontal handler would seem to make this method the most logical choice where the plant is concerned. A substantial cost savings due to higher quality (therefore, more good pieces per hour) would more than justify this new design.

Figure 7.6   Proposal to supervisor.

SUPERVISOR'S RESPONSE

To: Charles King, Junior QC Engineer
From: Tony Randall, Supervisor 1
Re: Tube Handler Proposal

Improved quality and simplicity of the horizontal design handler seem to be its strongest selling points. Its ability to adapt to new sizes readily gives it a definite advantage to the vertical-chuck type.

But multiple beam paths required for four beams and related shuttle systems would require an entirely new cutting head design. Quad optical system also would require a more critical alignment procedure.

Horizontal cutting would solve most of the costly problems now facing laser cutting the vertical way. The only advantage of the vertical method would seem to be that it uses much less floor space.

Figure 7.6 (*Continued*) Supervisor's response.

changes and explain how, and a second in which you dispute the instructor's interpretation of your material and defend your position.

b. If you hold or have held a part- or full-time job and have had to write any kind of report for which you received oral or written feedback,

- *list* and summarize the main points of the reader's response
- *write* two versions of a reply: one in which you agree with their suggestions or requests for more information, and one in which you do not.

The Process Applied

CHAPTER 8

Applying for a Job

There are three steps in preparing a good resumé: (1) make a list of everything you know about yourself; (2) analyze the list to decide what to include and what to emphasize, and (3) design a resumé in either outline or paragraph format that highlights either where you were educated and worked, when, your position titles, or your skills. The primary purpose of a resumé is to get you an interview.

Letters of application and interviews require the same reader/listener adaptation procedures covered in Chapters 1 through 3. Figure out what your perspective employer is looking for and stress those aspects of your accomplishments that most closely match your audience's expectations.

LISTING WHAT YOU KNOW ABOUT YOURSELF

One of the most important products that you will sell in written and oral communications is yourself. Preparing resumés, job application letters, and interview techniques involves the same before, during, and after procedures that you will use in all your technical writing tasks. Therefore, before you can present information about yourself to prospective employers, or admissions committees, you need to have a firm grasp of the facts.

As in the preparation of short technical reports, first list the details. Write down everything you can think of about yourself that is even remotely connected to your hireability. Put all of the information in front of you, so that you can sort and process it.

Here, for example, is a list prepared by a student a few months before he graduated from Case Western Reserve University with a degree in Electrical Engineering:

Main Interest: digital electronics, especially microprocessors and their use in industrial and consumer product development

EDUCATION

- CWRU 1975–79 B.S. in Electrical Engineering 5/79
- GPA, 3.25
- GPA in major, 3.5
- Areas of expertise
- Digital and analog design
- Microprocessor system designing and programming
- Courses in solid state electronics and psychology
- Member of Institute of Electrical and Electronic Engineers
- Dean's list 5 semesters
- Worked in Mechanical and Aerospace Engineering Department

EMPLOYMENT

- Mechanical and Aerospace Engineering Department of CWRU
  (Cleveland, Ohio 6/77–present) Worked 15 hours/wk., 40 hrs./wk. during summer
    months. Earned 20% of college expenses. Performed electronic design and
    repair. Projects included:

  - Assisting in the completion of a high-speed data acquisition network
    consisting of interlinked Z-80 microprocessors and one DEC PDP-11/40
    minicomputer.
  - Presently designing 16-channel Direct Memory Access (DMA) system to be
    included in the functional structure of the network.
  - Implemented a 4-channel DMA controller for use in Z-80 microcomputer
    systems.
  - Designed, built and tested an 8-channel analog multiplexer to be used by the
    Anesthesiology Department of University Hospitals in Cleveland.
  - Repaired and serviced existing hardware interface to a PDP-11/40
    minicomputer presently used for data acquisition.
  - Wrote software in 8080 Assembly Code necessary for testing and exercising
    various Z-80 peripheral boards designed and built as part of the network.

University Circle, Incorporated
(Cleveland, Ohio 9/75–5/77) 10 hrs./wk., 40 hrs./wk. during summer months.
General maintenance of buildings and facilities. Supervised four other
employees at day-to-day tasks. Earned 10% of college expenses.

Campus Mail Service of CWRU
(Cleveland, Ohio 5/75–8/75) 40 hrs./wk. Delivered mail to various buildings and
departments of the University. My first campus job.

St. Ignatius High School
(Cleveland, Ohio 5/71–5/75) 8 hrs./wk., 40 hrs./wk. during summer months.
Provided 90% of high school expenses. Duties included janitorial tasks,
dishwashing, tablesetting for resident Jesuits. In charge of two other employees.

PERSONAL

Born and raised in Cleveland, Ohio; interests include Astronomy, tropical fish
breeding, music, weight training, and racquetball.

DECIDING ON EMPHASIS AND FORMAT

Once the student had listed all his facts, he had to decide on a design for his resumé that, using these facts, would present him in the most favorable light to prospective employers.

As a writer in this situation, you always put yourself in your readers' place. Make a reader description list to remind you of the questions your readers will be asking. One general guideline is that a resumé should be in a format that the interested reader is accustomed to seeing. It should stand out from the 200 other resumés on the reader's desk because it is on unusually good paper or has an attractive (though not bizarre) typeface; or it should stand out, hopefully, because of the candidate's impressive accomplishments or potential. Your resumé can be creative, but it should not call attention to itself because it must be read from the bottom up or because it does not give the reader the basic components in a manageable pattern.

Basically, a resumé can be formatted as either an outline or a sequence of paragraphs. In either case, each section needs an appropriate heading, and the information is listed under each heading chronologically. You can determine which format is best for your situation in several ways:

1. The Placement Office or the department in which you are a major may keep a file of sample resumés for your profession.
2. Your college or local library will have books of sample resumés appropriate for different professions.
3. A professional society in your field may have an instruction pamphlet or guideline for preparing a *vita*.
4. A friend or relative who does hiring in your profession may be able to advise you.
5. Your housemates or Fraternity or Sorority may keep a file of sample resumés that work.

One or more of these sources will assist you in adapting the general format most appropriate for your profession or most familiar to the specific companies in your geographic location. Looking at sample resumés will help you decide whether to put educational background before or after work experience; whether and where to include a description of your hobbies; whether to write in outlines or in full paragraphs, and whether to write chronologically, beginning either with your current experience or with your earliest experience.

Whichever general format you choose, however, you will still need to custom design your material to meet two objectives:

• presenting yourself as favorably as possible without distorting the facts
• answering the questions in your readers' minds in such a way as to make them wish to interview you.

Although the resumé is usually sent along with a letter of application (or cover letter), you do not know if both will be read by all of the concerned parties and which will be read first. Your resumé, therefore, should be designed to speak for itself and should not rely on the letter of application to precede it.

What will your reader look for? Most potential employers want to know what special training, experience, or aptitude makes you particularly well qualified for the job they are offering. If a special qualification is claimed, your readers want to know what substantiates the claim. If your resumé demonstrates that you have a special qualification, you are very likely to be granted an interview.

DESIGNING THE RESUMÉ

If you have a clear idea of the kind of job you want, you can prepare a basic resumé that will probably require only occasional updating. If you have different skills that qualify you for more than one type of position or company, you will want to develop more than one standard resumé, one for an academic job and another for a corporate position, for example. If possible, create a separate file on your word processor for each version of your resumé.

While the basic components of a resumé are essentially the same: Education, Experience, Outside Interests, and Recommendations, formatting of facts within each category can be arranged and re-arranged to achieve different emphases. Under "Experience," or "Employment History," for example, you must include dates of employment, company's name, job title, and job description. But which of these you highlight and which you subordinate should depend on which structure will present you most favorably.

For instance, if you have worked for three prestigious firms in your field but have only held apprenticeship or junior positions, you may want to stress *who* you worked for rather than for how long or what you did.

For example, an accounting major has held jobs at three organizations that are well-known in the Midwest. She worked only during the summer months, and only one of the positions held is directly related to accounting. Using the outline format, she would want to set up the part of her resumé dealing with employment history as follows:

EMPLOYMENT

Union Carbide Corporation
   Ferroalloys Division; June–August 1986; tapper

Case Western Reserve University
   Systems Engineering Department, June–August 1985; lab assistant

Ashtabula Rubber Company
   June–August 1984; utility person

In contrast, the electrical engineering student in our earlier example has held three jobs, each with a different division of the same university, and one at his high school. His strong point is not where he worked, but the skills that his job experience demonstrates. Therefore, he would want to stress *what* he did, rather than where:

EMPLOYMENT

Electronic Design and Repair
  June, 1986–present: Mechanical and Aerospace Engineering Department, Case Western Reserve University
Supervision and Maintenance
  September, 1984–May, 1986; University Circle, Inc.
Mail Delivery
  May–August 1984; Campus Mail Service, Case Western Reserve University

As you advance in your career, you may want to stress the length of time you were at your previous jobs to demonstrate reliability or to emphasize the lack of "gaps" in your employment record. If so, you would highlight the dates of your employment:

EXPERIENCE

October, 1986–present: programming supervisor, Xerox Corporation.
Responsibilities:. . . .

May, 1984–October, 1986: junior programmer, Xerox Corporation.
Responsibilities:. . . .
June 1975–May 1977: junior programmer, IBM Corporation. Responsibilities. . . .

If you are a woman re-entering the job market after years as a homemaker, or a person who has switched jobs frequently or been out of work often, you must be prepared to explain these situations comfortably at an interview. Nevertheless, you would not want to call undo attention to time gaps in your resumé. If this is ever your situation, highlight your skills or your employers, or something other than the dates of your employment. If you have worked at two or more companies at the same level and want to stress to your perspective employer that you have enough experience to be hired at the next level up, you may want to highlight job title on your resumé:

EXPERIENCE

Junior programmer I
  June, 1982–May, 1984: IBM Corporation. Responsibilities:. . . .
Junior programmer II
  May, 1984–October 1986: Xerox Corporation. Responsibilities:. . . .
Junior programmer II and Assistant to Supervisor of Programming
  October, 1986–present: Xerox Corporation. Responsibilities:. . . .

The difference that proper highlighting and design can make is demonstrated by the original and revised versions of a management major's resumé shown in

Figures 8.1 and 8.2, respectively. Since the student is applying for a management position, he wants to stress his leadership qualities. In the first version, it is up to the reader to find the leadership qualities which are somewhere in the information. In the revised version, the qualities most likely to get this senior his entry-level job in management stand out.

Kurt D. Warrington

653 Apple Rd   12 Thorndike Rd.
Cleveland, Ohio 44106   Maplewood, NJ 07112
(216) 555-3837   (201) 443-1218

EDUCATION

Case Western Reserve University, August, 1982 to May, 1986
    Bachelor of Science Degree in Management (Minor in Finance)
    Along with my required courses, I have taken courses to supplement and round
    out my minor.
        Examples: Tax Accounting Computer Forecasting,
        Budgeting and Simulation and
        Business and Technical Writing

ACTIVITIES AND AWARDS

Case Western Reserve University Circle Chorale (U.C.C.), 4 school years
Co-chairman of U.C.C. student organization, 1985 to 1986
Society for Advancement of Management (S.A.M.) of the American Management
    Association
S.A.M. Liaison to National and Summer Jobs Program Coordinator, 1983 to 1985
S.A.M. Vice President of Professional Activities, 1985 to 1986
Guest Lecturer at Pequannock Township High School, 1983 to present
Recipient of S.A.M. Membership Scholarship, Summer 1986

JOBS HELD

Weekend Custodian, Church of the Covenant, Spring 1981
                    Euclid Avenue
                    Cleveland, OH 44106
Control Room Operator, Instructional Television Network—1982 to 1984
                    Case Western Reserve University
                    Cleveland, OH 44106

REFERENCES

Available on Request

Figure 8.1 Original resumé.

Kurt D. Warrington

653 Apple Rd
Cleveland, Ohio 44106
(216) 555-3837

12 Thorndike Rd.
Maplewood, NJ 07112
(201) 443-1218

OBJECTIVE: Within three to five years, be in a middle-management position and be able to train others along the same track.

EDUCATION

Case Western Reserve University, August 1982 to May 1986
B.S. in Management (specialization in finance)
Supplemental courses in tax accounting; computer forecasting, budgeting, and simulation; and business and technical writing.

MANAGEMENT/ORGANIZATIONAL EXPERIENCE

Vice-President of Professional Activities: Society for the Advancement of Management (S.A.M.), 1982–1983
Duties: Responsible for all professional activities of the group
    Program planning and coordination
    Scheduling of all professional speakers
    Selling the organization to the local business community

Summer Job Program Coordinator: S.A.M., Spring 1982. Initiated the idea for the program.
Duties: Develop the program
    Write and send letters to 80 Cleveland-area businesses
    Process all responses

Liaison to the American Management Association for S.A.M., 1981 to 1982
Duties: Coordinate and oversee the securing of S.A.M.'s charter from the A.M.A.
    Sustain correspondence with the A.M.A.
    Maintain points recognition program

Co-Chairman, University Circle Chorale
Duties: Set up organizational structure
    Draft a working constitution

WORK EXPERIENCE

Church of the Covenant, Euclid Avenue, Cleveland, OH 44106: Weekend Custodian, Spring 1984
Instructional Television Network, Case Western Reserve University, Cleveland, OH 44106: Control Room Operator, 1982–1983
Warrington's Photographic Services, Maplewood, NJ 07112: Self-owned and -operated business, 1980–1982

RELATED ACTIVITIES

Society for the Advancement of Management: two school years
University Circle Chorale: four school years
Guest Lecturer on small business starts, employee/employer rights: 1982 to present
Recipient of S.A.M. Membership Scholarship: Summer 1983

References
Available upon request.

Figure 8.2 Revised resumé.

WRITING LETTERS OF APPLICATION

When you send a resumé, accompany it with a cover letter. Since you do not know which will be read first or which will be more important, the letter should indicate briefly the same highlights of your qualifications for the job that are represented in your resume.

Solicited Application Letters

The firms that advertise in a newspaper or professional journal or send recruitors to your college or university are announcing specific jobs available for which you may qualify. They are soliciting letters of application, and when you write yours, you will usually have their published job description as a guideline for the questions they want you to answer about your qualifications.

Example 1

New England Nuclear advertised for chemical technologists in the *Boston Globe,* March 25, 1978 (Figure 8.3). Figure 8.4 is a letter of application for this position written by a student about to receive a combined B.A. and M.S. degree in Biology.

Notice that the first sentence of the letter makes clear which position is being applied for and the writer's source of information. This is especially important when responding to an ad for more than one position. The second sentence of *paragraph one* indicates the applicant's educational qualifications and when he can begin work. *Paragraph two* indicates that the applicant has the experience needed for the job. *Paragraph three* subtly eases the readers' minds about the possible difficulties a young college graduate may have adapting to a new location. This applicant will not have such difficulties because he has lived in Boston before.

Most important of all, this letter of application is short, to the point, answers the readers' questions, and makes them want to know more. They will read the attached resumé, and if it confirms the impression made by the letter, they will call the applicant for an interview.

Example 2

Figure 8.5 is another example of a letter in reply to a specific job advertisement. In this letter, the applicant is dealing with the problem we all face at the beginning of our careers: he has lots of enthusiasm but little or no experience. (Sometimes it seems as if everyone wants beginners with experience, but no one is willing to give beginners their first experience.) In trying to compensate for lack of experience, this applicant falls into the trap of overkill. By continually reminding the company of what he lacks, he makes his inexperience seem more of a liability than it really is. Also, given the information the letter has to impart, it is far too long.

ANALYTICAL OPPORTUNITIES

New England Nuclear, an internationally recognized leader in the manufacture of tracer chemicals for medical diagnosis and research, has several new opportunities for a wide range of experienced professionals seeking exciting and challenging opportunities. The rapid expansion of our product lines has created an immediate need for our Corporate Analytical Services Department to develop and refine analytical methods and procedures to meet the demands of this new growth. Our Analytical Services Department provides support to our facilities in Boston, Billerica and Westwood.

Analytical Chemists

Boston & Westwood

Your primary responsibility will be the development of new methods of analyses with strong emphasis in the area of GLC & HPLC. To qualify for these positions you should possess a B.S. in Chemistry and have 0-2 years experience with analytical instrumentation.

Chemical Technologists

Boston & Billerica

We have several positions available for individuals with a B.S. in a life science and experience with analytical instrumentation and chromatographic methods of analysis.

Methods Development

Boston & Billerica

Assist in the development of new methods of analysis, modification, and improvement of existing methods.

Site Supervisor Corporate Analytical Services

Billerica

Ph.D. in Chemistry, Life Sciences or Physics with a minimum of 1-2 years industrial or post-doctoral experience with strong supervisory experience.

Within your background a strong knowledge and exposure to conventional analytical and nuclear instrumentation will be expected. Some exposure to programming in BASIC and FORTRAN is essential.

You will be responsible for the initial establishment and coordination of this effort within our Billerica site.

In-Process

Boston

Assist in the development of critical analyses in evaluation of materials from production labs. Communicate with chemists to insure that products meet specifications.

Quality Control

Boston

Emphasis will be placed on adapting GLC & HPLC methods to replace current PC & TLC methods.

Forward resumes to the attention of Ann Magno.

Programmer Analyst

Boston

B.S./M.S. in a Science with a minimum of 2 years programming and analysis experience in FORTRAN and BASIC in the areas of chemistry, or the life or physical sciences. Previous experience in a laboratory environment and familiarity with analytical and nuclear instrumentation is needed.

You will be asked to create and develop computer programs for our production oriented requirements.

NMR Spectroscopist

Boston

B.S. in Chemistry with a minimum of one year on-hands experience in NMR operation (iron core and supercon). Your background will be applied to the daily operation plus care and maintenance of our NMR facility.

Forward resumes to the attention of Janet Anderson.

In return for your expertise, NEN offers competitive compensation, an informal yet highly professional environment and a top benefits package including Blue Cross/Blue Shield and Profit Sharing. Interested applicants may forward their resumes with salary history to:

New England Nuclear
Personnel Department
549 Albany Street
Boston, MA 02118

(NEN) New England Nuclear

An Equal Opportunity Employer

Figure 8.3 Job advertisement.

1745 E. 116 Place #44
Cleveland, OH 44106
April 20, 1978

New England Nuclear
Personnel Office
549 Albany Street
Boston, MA 02118

Dear Sirs:

I would like to apply for the position of chemical technologist in the Methods
Development Department as advertised in the March 25, 1978, *Boston Globe*. I
will be able to start June 1, 1978, following my graduation with a Bachelor of
Arts degree in Biology and Chemistry and a Master of Science degree in Biology
from Case Institute of Technology.

I have two and one-half years of experience working in research laboratories
using several types of analytic instrumentation and chromatographic methods.
I have adapted several procedures from the literature for use in the laboratory.

I would like to return to the Boston area and look forward to interviewing with
you.

Very truly yours,

Stanley Cutter

Figure 8.4   Solicited letter of application.

Figure 8.6 is a revised version, which maximizes facts and minimizes explanations. The letter emphasizes the experience the applicant has, not what he
does not have. Remember, the purpose of the letter and resumé is to make the
readers want to talk with you further about employing you. Save detailed explanations for the interview itself.

Unsolicited Applications

Sometimes, it is to your advantage to call an employer's attention to your existence even when no specific job exists at present. If you identify a need that you
can fill, a position may be created for you. Or, when a regular position suited
to your qualifications does open up, the company may pull your application
from its active file. (If you do send an application that will not be acted on immediately, be sure to send the company an updated resumé periodically.)

11205 Bellflower Rd.
Cleveland, Ohio 44106
March 3, 1987

Cleveland-Cliffs Iron Co.
1460 Union Commerce Bldg.
Cleveland, Ohio 44115

Dear Sir:

Through the 1986–87 College Placement Annual, I found that the Cleveland-Cliffs Iron Company hires accounting students for summer employment. I would like to be considered for a position in the Cleveland office.

Presently, I am completing my junior year in accounting at Case Western Reserve University. A strong background in accounting and finance should qualify me for a management position.

Although my previous work experience of this nature is limited, I think I can overcome this factor with enthusiasm. I am very excited about having an opportunity to apply my education and perform in a work environment. Also, my versatility should help make up for a lack of job experience. As an example, I think I have done well at balancing the tradeoffs between successful academics, work-study employment, responsible fraternity membership, and recreational activities.

I am sure that I could be a definite asset to Cleveland-Cliffs for the summer, and possibly during the school year on a part-time basis.

I am looking forward to an interview and could begin work on May 19, 1987. Thank you for your consideration.

Sincerely,

Figure 8.5   Weak cover letter.

Our electrical engineering applicant was able, for example, through the help of relatives and friends, to locate several firms that could use his special expertise and that were located in areas where he would be willing to relocate. Figure 8.7 is a letter he wrote to the division supervisor of a company where one of his friends holds a position.

Whenever possible, refer to a personal contact known to the company when sending an unsolicited application. Indicate specific knowledge of what the company does and why that especially interests you. Then state your qualifications

                                                          11205 Bellflower Rd.
                                                          Cleveland, Ohio 44106
                                                          March 3, 1987

Cleveland-Cliffs Iron Co.
1460 Union Commerce Bldg.
Cleveland, Ohio 44115

Dear Sir:

Through the 1986–87 College Placement Annual, I found that the Cleveland-
Cliffs Iron Company hires accounting students for summer employment. I
would like to be considered for a position in the Cleveland office.

Presently, I am completing my junior year in accounting at Case Western
Reserve University. I have taken 13 courses (39 credits) in accounting and
finance and have a 3.62 average.

During previous summers, I held positions of responsibility at Union Carbide
Corporation, Case Western Reserve University, and Ashtabula Rubber Company.
I can offer a combination of education, job experience and enthusiasm that
would be an asset to your organization.

I could begin work on May 19th, and would be available full-time through
August, and part-time afterwards. I'd be pleased to arrange an interview at
your convenience.

                                      Sincerely,

Figure 8.6   Strong cover letter.

as briefly and precisely as possible; indicate how and when you can be reached
and end your letter graciously. Short, to the point letters are always appreciated.

THE JOB INTERVIEW

Think of a job interview as an extemporaneous oral presentation. If you arrange
for an interview with a company recruiter or a private employment agency
through your college placement office, this oral presentation may be the company's first contact with you. If you are called for an interview in response to a
written application, then you know that you have survived the initial sorting-out
process and that the firm is seriously interested in you. Either way, your purpose
in the interview is to make a good first impression or to confirm and broaden
the positive impression the potential employer has already.

14233 Triskett Road #301
Cleveland, Ohio 44111
April 30, 1979

Mr. Roger Olson
Mail Drop B36
P.O. Box 2000
Phoenix, Arizona 85005

Dear Mr. Olson:

I will receive a degree in Electrical Engineering from Case Western Reserve
University this May, and have a 3.5 grade point average in my major. Walter
Anderson, an engineer working in your division, has described to me in some
detail the types of job assignments available and the possible opportunity to
pursue an advanced degree in engineering. I am able to relocate to Phoenix, and
would like to discuss a career possibility with you.

At present I am completing the design of a data acquisition network utilizing
several interfaced Z-80 microcomputers and one DEC PDP-11/40 minicomputer. I
have found this job experience instrumental in reaffirming my interests and
desired career goals in computer technology.

My resume is enclosed for your convenience. Please note that I successfully
financed my own education for eight years, combining work and education.
Further information is available at your request.

I hope to hear from you soon so that you may get to know my interests and
background better. You can reach me evenings at 217-221-1338 or at the above
address.

I appreciate your time and consideration.

Sincerely,

Figure 8.7   Unsolicited letter of application.

What Are the Interviewers Looking For?

When your prospective employers interview you, they are trying to judge how
you will conduct yourself in a business situation, how you will represent their
company:

- Do you interact well with others?
- Do you express your ideas simply and clearly?

- Do you have decisive opinions on issues?
- Can you accept criticism?

Unless they are hiring you on a temporary basis (in summer or part-time), they will also be thinking in terms of your continuing value—not only what you can do now, but what you might be able to do five years from now. They will want to know what you see yourself doing over the next few years, what career track you want to follow, what new skills you might develop, and so on. Be realistic in answering these questions.

Preparing for an Interview

As with any other written or oral communication, the most effective preparation for an interview is to (a) know your material, and (b) know your audience.

Know Your Material

You do not have to memorize the address of your high school or be able to recite your resumé by heart. In fact, a canned performance of memorized facts will make the interviewer wonder if you are capable of understanding things without memorizing them. What you do need to have clearly in mind are the key features of your education and experience. If you know them well, you can bring them into play in answering the interviewer's questions.

You also need to know about any gaps in time or skills that your resumé might reveal, so that you can have an explanation ready. Even if you were hitch-hiking through Switzerland the year you were out of work, say so matter-of-factly. Never be apologetic or humble in responding to a reasonable question. Be calm and decisive in everything you say. Do not sound as if you had rehearsed everything, even if you have to rehearse in order to sound like you have not.

Know Your Audience

Find out as much as possible beforehand about who will interview you. If you know people they have interviewed previously, contact them and find out what kinds of questions they were asked. If one or more of your interviewers has published, familiarize yourself with his/her work.

Ask your placement office or library for sources of information on the company that will interview you, such as its annual report. One of the questions the company will ask you is what you know about them, and why you want to work for them.

Practice Your Style

Going into an interview "cold turkey" without any previous experience can be a rude awakening. It is best to find out how you behave in this type of situation *before* you face the real thing.

You might, for example, make up a list of hypothetical questions and give it to two or three friends who are also preparing for interviews. Take turns playing the role of interviewer and interviewee. If possible, use an office or a part of a room that can be arranged like an office.

As the person to be interviewed, practice walking into the room decisively, body thrust slightly forward, smiling and shaking hands firmly with all those present. Be sure to look each person straight in the eye. Direct eye contact helps them focus on what you are saying and gives the impression of self-confidence. Also, if you actually look at the people interviewing you, you will discover that they are a lot less frightening than they appeared in your imagination. In your zealousness to show confidence, however, avoid taking over. Your interviewers are in charge and will more or less determine what is covered when, and how long the interview lasts.

Control in an Interview

The most important way you can control an interview is by being relaxed. When you relax, everyone else does. Their questions become more interesting, and so do your answers. If, on the other hand, your interviewers sense your nervousness, they will concentrate on making you feel relaxed. You may wind up talking about golf or bridge or the weather for much longer than anyone present desires. It is up to you to signal by your behavior that you are relaxed and ready to get down to business.

Fielding Questions and Asking Them

Have your friends throw in some unexpected questions in the mock interview. No matter how carefully you have investigated your interviewers, no matter how many interviews you have experienced, you will almost always be asked at least one question you did not anticipate. Try to practice not appearing startled or thrown by the unexpected. A comment such as "I might be able to answer that question more accurately if you'd tell me a little about why you're asking it" is perfectly acceptable. You don't want to overuse this technique of putting the ball back in the interviewer's court, but you do want to let your audience know when you need more information from them to answer their questions.

Remember, the purpose of practice or mock interviews is not to prepare canned speeches. These will work against you rather than for you. Instead, the purpose of practicing is to develop a style that projects confidence, to make yourself comfortable with the interview dynamic, so that you can react positively and creatively to whatever happens at the real interview.

Do not answer positively unless you mean it. If, for example, you do not want to go on for an advanced degree, say so and explain why. Do not worry about trying to fit what you think their image of you may be. A little independence is a positive trait. Besides, if you really are very different from most of the organization's personnel, you might as well all know that from the beginning. Unless you like being a square peg in a round hole, there are some jobs you would be better off not getting.

Aside from answering questions, you will want to learn how to ask questions. Discussions of salary and benefits are better left to private consultations with one personnel officer or the appropriate manager. But at the general interview, you will want to initiate questions about how the firm operates and how you will fit in, and about your specific responsibilities and the kind of backup you will receive. Asking questions about the everyday workings of the company and about your responsibilities and how you will fulfill them demonstrates your knowledge of the working world and your real interest in doing a good job.

If you conduct your mock interview with at least two different sets of friends or relatives, so much the better. An important part of successful interviewing is learning how to respond well to different personalities.

FOLLOWUP LETTERS

Before you leave an interview, try to agree on a specific time by which the company will contact you again. If, for example, they agree to call you in two weeks, you at least know how long your anxiety has to last. You also know when a followup letter will seem reasonable to all parties concerned. At the end of two weeks, if you have not heard from them, you can write or call to remind them

Figure 8.8 Followup letter.

June 4, 1987

Ms. Eleanor Jordan
R & D Administrator
Barton Enterprises
Seattle, Washington

Dear Ms. Jordan:

I am now back to the academic grind in Boston, but I wanted to take a minute to thank you and Mr. Jordan for your gracious hospitality during my visit to Seattle for an interview with your firm.

Since you seemed interested in the microprocessing design variations that I was working on with Prof. Minton, I requested and received permission to send you sample specs. The drawings, of course, are under copyright protection, and you would need Prof. Minton's permission to make further use of them.

Thank you again for your time and interest. After talking with you and your staff, I feel that I could make a valuable contribution to Barton Enterprises and hope that you will decide to give me the opportunity.

Sincerely,

Jim Young

politely that you are waiting to hear. However, especially if they will not be contacting you for at least a week, you may want to send a followup letter just to say that you enjoyed meeting them and are really interested in working for them. It takes practice to make this statement sincerely and without sounding over-eager. If possible, find an incidental occasion for writing. For example,

- an article you had read came up in the discussion, and you promised to send one of the interviewers the exact citation or an off-print.
- a piece of equipment that you just found out about would be useful in a project their company is undertaking.
- you will be at your parents' cottage on the Cape for the next two weeks and wanted to let them know where they could reach you.

If the interview involved an overnight or weekend visit, you can write to thank them for their hospitality. A sample followup letter appears in Figure 8.8.

ACCEPTANCE AND NON-ACCEPTANCE LETTERS

If you are offered the job and decide to accept, respond promptly and precisely. Even if the offer is made verbally, a written confirmation will be sent outlining

Figure 8.9 Acceptance letter.

May 10, 1987

Mr. Roger Martin
Communications Division
XYZ Corporation
Phoenix, Arizona

Dear Mr. Martin:

I am pleased to accept your offer of a position as junior technical writer at an initial salary of $21,000 per year, effective June 1, 1987.

I understand that my appointment in the Phoenix Office is for two years subject to bi-annual reviews, and that I may at the end of two years request a transfer to the San Francisco office.

I look forward to working with you and your staff as a full-time member of the team.

Sincerely,

Bill Snider

---

title, length of contract, salary, responsibilities, and so on. Repeat these details in your acceptance letter, so that there is no question that both the employer and the employee have the same understanding of the agreement that they are making (Figure 8.9).

If you are offered the job and decide not to accept it, write a courteous reply, leaving the door open for the future. Thank them for their interest. If you have accepted another job instead of theirs, tell them; but phrase your explanation in terms of how your interests for the present are better served by the other company, rather than in terms of how the other company is better than theirs. Remember that you may be rejecting a future employer or customer (Figure 8.10)

Figure 8.10   Non-acceptance letter.

                                                                                     April 2, 1987

Mr. Roger Martin
Communications Division
XYZ Corporation
Phoenix, Arizona

Dear Mr. Martin:

I have given careful consideration to your offer of a full-time position as a technical writer when I graduate in June, 1981.

As you know, this summer's internship with XYZ Corporation was a practical learning experience for me, and I value the high regard that has prompted your offer.

While I appreciate the advantages of beginning my career with a company such as yours and the opportunity for growth, I must weigh these facts against my original plans to attend law school immediately following graduation.

Because it is in the best interests of both myself and my future employers for me to increase and develop my academic skills before getting side-tracked by the responsibilities of a full-time position, I must reluctantly refuse your offer.

I am grateful to you for your interest in me and hope that there will be an opportunity in the future for us to work together.

                                                    Sincerely,

                                                    Bill Snider

---

EXERCISES

1. *Resumés*—Make a list of all your interests, skills, achievements, experiences—anything and everything that might make you attractive to a potential employer. Analyze the list to determine which items to include on your resumé and in which order; then write an effective resumé based on your decisions.

2. *Job Application Letters*—Find an ad in the newspaper or in a professional job listing that describes a position for which you are qualified presently or for which you will be qualified when you complete your formal education. Write a letter of application for the position advertised, designed to get you an interview.

3. *Interviews*—Conduct a series of mock interviews in your class based on real jobs listed by your placement office. Analyze each candidate's performance. Which questions presented problems and why.

   If possible, have local business personnel or Placement Officers conduct a mock interview for your class. Analyze the kinds of questions they ask, and make sure you understand the purpose of each question.

Writing Letters, Memos, and Minutes

Writing several short technical communications to persons inside and outside your organization will be part of your daily routine. Short memos and letters usually (1) give or request information, (2) complain or respond to complaints, (3) persuade or respond to persuasion. Each instance requires the writer to "sell" the information to the reader.

1. Giving or Requesting Information
   a. Make the point clear from the beginning in as few words as possible.
   b. Use headings to separate parts.
   c. Use list format instead of paragraphs where possible.

2. Complaining
   a. Describe the product or situation.
   b. Describe what you want done.
   c. Give the reader a deadline.

   Responding to Complaints

Good News	*Bad News*
a. Tell the reader exactly what you will do.	a. Present facts/reasons *before* refusing.
b. Give necessary dates, instructions, and so on.	b. Offer a compromise or alternative.
	c. Close on a positive note.

3. Persuading or Responding to Persuasion
   a. List the facts.
   b. Determine the priority message.

    c. Anticipate the reader's perspective.

    d. Write a letter/memo designed to convince that particular reader of the priority message and supporting facts.

GIVING OR REQUESTING INFORMATION

Unsolicited Information

Often you will need to notify peers, subordinates, or supervisors of changes in conditions, procedures, schedules, personnel, or equipment. Especially if several people are affected, you will probably write a general-information memo, or release to the entire staff a copy of the explanatory memo you sent to your boss.

Usually, the purpose of your communication is twofold:

- to enable those working with you and under you to do their jobs efficiently
- to let those to whom you are responsible know that you are doing your job efficiently.

If you can imagine the number of memos of this kind that each staffperson could conceivably have occasion to write per week, you will realize that the information you have to convey is in competition with stacks of other memos asking the same readers to absorb and hopefully retain their messages. If you want a successful communication transfer to take place, the burden is on you to make it as simple as possible for the reader to get your point. Your letter or memo should:

1. Make the subject clear from the beginning.
2. State the point in as few words as possible.
3. Use straightforward headings to identify different or subordinate points quickly and easily.
4. Use a list format instead of a paragraph format where appropriate.

If how the particular readers are affected by the new information is not self-evident, tell them what they will do or experience differently as a result of your communication.

Example 1

Figure 9.1 is a memo issued by Janet Lund, a field service administrator for a company that sells medical test equipment, regarding a new program for controlling inventory of recorder parts. The statement of the subject clearly identifies the nature of the information so that readers affected by this subject will know that they should give this memo their attention. The arranging and under-

To   Paul Lewis   From Janet Lund
Date   March 27, 1987
Subject   Program for Surplus Run of Field and   Copies to R. Meltzer
          Obsolete Inventory   D. Torelli

As a result of my request for a surplus run of field and obsolete inventory, D.
Torelli will write a new program for the field inventory usage. This run will
display for each of the branch inventories and a total of all the branch
inventories:

- One year past usage
- Stock on hand
- Excess of stock on hand.

This run will enable each of the branches to:

1. Forecast and order parts they know they need. With the current delivery
   from Cleveland, Service should be able to reduce their inventory level from
   18 months (Jan. 1987) to six months or a total projected decrease in existing
   inventory of ± 350,000.
2. Get rid of excess inventory, either by returning it to Cleveland or
   obsolescence.

The run of the total branch usage will enable us to establish depot stock (Los
Angeles, Cleveland, Saddlebrook) for items that are critical, low usage and high
dollar parts.

Example

Part #	Description	Acct.	Usage	Qty. on Hand
778	Drive Board	031461	12	12
		031468	14	10
		031476	24	2
		031483	49	18
		Total	99	42

Figure 9.1 Effective memo.

lining of the key material makes it easy for the reader to see the three main
points quickly:

1. The new program will show one year's past usage, stock on hand, and excess
   stock on hand for each of the branch inventories and a total of all the
   branches.
2. By using the run, each branch can forecast and order ahead the parts it will
   need and get rid of those it does not need.
3. Having a run of total branch usage enables the establishment of a depot stock
   for low-usage/high-dollar parts.

Sometimes, the information you need to impart is not the announcement of a change but of the need for a change. You will have been aware of some aspect of your operation that is inefficient or less efficient than it should be. You have determined the appropriate party to refer this matter to and now you need to write an explanation.

Example 2

The same field administrator who wrote the previous example was having difficulty writing a memo to explain the EDC Scopes dilemma. She presented a copy of her first attempt (Figure 9.2) at a technical writing seminar. Her fellow students agreed that it was difficult to understand her point. The subject announcement indicated only that the memo was about service parts for EDC Scopes, but it did not indicate the specific message. The physical appearance of the body, several short paragraphs with no headings, did not enable the readers to get a quick picture of how many points there were, or of what they were.

Figure 9.2   Memo: first draft.

To	E. Walters	From Janet Lund	

Date   September 21, 1979

Subject   Service Parts for Marcom Scopes   Copies to   B. Morris
                                                                     T. Sargent
                                                                     J. Keene
                                                                     M. Winkler

The material in 012864 to support the EDC scopes was a projection (forecast) by Service based on the scope forecast by Marketing.

Though these parts are not required to manufacture a product we sell, they are parts to support a product we sell.

If a distinction is being made by Production Control between equipment we support and equipment we manufacture, a separate home account inventory must be maintained for the "support" parts. Our present House Order System allows only for the recognition of a home account stock-on-hand. If no stock-on-hand exists a backorder is created, which elicits a demand to order.

Don Sears has advised me that the only other account these parts could be moved into is 0313662; however, this is a recognized obsolete inventory account and should not be used for current items.

This is not the only "third" party equipment we are supporting but not manufacturing. A management decision must be made as to how and where these parts can be maintained that will fit our House Order System.

It was suggested that the writer list all of her facts, determine a priority message (as discussed in Chapter 4), and structure her memo by deciding whether it fit the comparison, problem/solution, or description format best. The list of facts (Figure 9.3) enabled the writer to decide that her priority message was the need for a management decision on separate storage of third-party parts within the existing inventory system. The best way to present these facts for this purpose was to use a problem/solution format. Her revised memo (Figure 9.4) is easier to read because it identifies the subject up front and gives the reader the means for grasping the main points quickly and easily.

Informing as a Means of Persuading

Sometimes, the information memo is also a request for action. Its readers must not only absorb the explanation, but realize what is being asked of them. Sometimes, you are explaining something to them, not only to get action, but to get action by a specified deadline.

The memo in Figure 9.5 is written by a member of the International Sales staff of a manufacturer of instruments for medical tests addressing the Director of Operations. All the reader knows about the subject designated is that it somehow concerns two orders for equipment. Casting his or her eye over the six unlabeled paragraphs, the reader has no way of grasping the main points quickly. In fact, however, the memo contains an urgent message and a deadline. An important sale could be delayed or even lost if the reader does not understand what is needed and when.

Figure 9.3   List of facts.

SERVICE PARTS FOR EDC SCOPES

- The material in acct. # 012864 to support the EDC Scopes was a projection by Service based on the Scope forecast by Marketing.
- These parts are not required to manufacture a product we sell, but to support it.
- If a distinction is being made by PC between equipment we support and equipment we manufacture, we need a separate home account inventory for support parts.
- Our present HOS allows only for recognition of stock on hand. If there's no s-o-h, a back order is created, which elicits a demand to order (inappropriate for support stock).
- In our present HOS, the only account these parts could be moved to is 0313662 which is also inappropriate because it is a recognized obsolete-inventory account and should not be used for current items.
- What is needed is a management decision on how and where to maintain parts for 3rd-party equipment within our existing HOS.

Subject: How to stock support parts for EDC Scopes and other products

PROBLEM

The service parts for EDC Scopes (acct. 012864), like some of the other equipment we stock, are required, not to manufacture a product we sell, but only to support our products. The EDC Scope parts and support parts generally are not being inventoried efficiently because our present House Order System has no account that accurately accommodates them.

   Account #012864, like all our home accounts, allows only for the recognition of a home account stock on hand. If no stock on hand exists, a back order is created, which elicits a demand to order.

   Account #0313662, according to Don Sears, is the only other place for the EDC scope service parts. But because it is a recognized obsolete inventory account, using it for current items would cause confusion.

SOLUTION

Management needs to decide how and where to maintain support parts, such as those for the EDC Scopes, within our present House Order System. If it is the policy of Production Control to make a distinction between equipment we support and equipment we manufacture, perhaps the best solution is *to establish a separate home account inventory for support parts only.*

Figure 9.4   Revised memo.

If readers have several pieces of correspondence to go through, most will read and act on the ones that look important first. Many professionals sort the materials in their "in" box into A, B and C stacks. "A" gets immediate attention; "B" gets looked at if time allows; and "C". . . .

   If you want to be in the "A" category, the responsibility is yours to make the importance of your communication evident. This author, with the aid of a technical writing instructor, re-wrote his memo (Figure 9.6), to make its priority more apparent.

Solicited Information

Often, your informative letter or memo will be in direct response to a specific request. If the request is in writing, you have the advantage of using the information provided by the requestor in formulating your reply. If the request comes verbally, try to take notes on what was asked for, either during or immediately after you speak with the intended readers.

   A Lake Electric engineer wrote the information memo in Figure 9.7 in response to a telephone conversation with his boss. He had been given the

To: Steve Rayburn   From   Ed Norton
Date: April 12, 1985
Subject: HOUSE ORDER No. 23-7654 and   Copies to S. H. Jones
23-76321   E. T. Morgan
                                                           G. R. Taylor

The above house order numbers are for the certification 5000 series recorder systems for Japan. It is with utmost urgency that we configure this system for delivery the first week of May.

The reason for the urgency is that negotiations with the Ministry of Health and Welfare of Japan and the associated Doctor Suzuki of Kyoto Memorial Hospital have scheduled their time for the end of May. This is of the utmost importance since these people from the Health and Welfare Ministry must certify the system before any recorders can be shipped to Japan.

A minimum of 30 to 60 live patient tests must be conducted on our recorder prior to September, 1985. However, the paperwork that is involved by the Health and Welfare Ministry will be conducted at the same time the valid tests are going on, with periodic spot checks by them.

Also, to educate Doctor Suzuki and the Kyoto surgical staff, I am scheduling my attendance in Kyoto to coincide with the clearance of customs near the mid or end of May. At all costs, the equipment should be in place prior to June 1.

We would appreciate your help in expediting this system so that we can take advantage of the U.S. Trade Center show interest prior to the fourth quarter of this year.

If we do not, there is a good chance that we can lose another 6–12 months in securing the necessary certification by the Ministry people because Doctor Suzuki will be leaving for a six-month sabbatical in the U.S. starting around mid-September of this year.

Figure 9.5   Urgent memo: first draft.

responsibility of getting bids for equipment Lake Electric needed to complete a project and for reporting on the choices available. His boss felt that a reasonable amount of time had passed and wanted an explanation of where they stood.

The memo is in effect a progress report in which the writer gives a play-by-play account of what he has done to get the bids and what the results are to date. The information is all valid and relevant, but the way the memo is set up, it is difficult for the reader to get the main point.

In a technical writing seminar at the electric company, the participants pointed out the need for a summary statement preceding the details. Following

---

URGENT!

House order nos. 23-7654 and 23-76321, the certification series recorder systems for Japan, *must* be assembled for delivery by the *first week in May.*

Explanation of Problem

Before Gould biophysical recorders can be imported by Japan, the Health and Welfare Ministry (HWM) must certify the system. We are scheduled to meet with HWM and Dr. Suzuki to receive this certification at the end of May.

    If we do not have the necessary equipment in place *before June 1st,* we may lose another 6 to 12 months before securing the necessary certificate, since Dr. Suzuki will be leaving Japan in mid-September for a sabbatical.

Requirements of the Certification Process

A minimum of 30 to 60 live patient tests must be conducted on our recorder prior to September, 1985. To help expedite this process, the HWM has agreed to do the necessary paperwork while the validation tests are in progress and to make periodic spot checks.

To make sure that the equipment clears customs smoothly, I am scheduling my visit to Kyoto to coincide with delivery.

What We Need from Operations

We are relying on you to have the equipment assembled for shipment in time to clear customs and reach Kyoto before the end of May.

Goal

With your cooperation, we can capitalize on the interest generated in our recorder at the U.S. Trade Center Show *before* the fourth quarter of this year.

Figure 9.6   Urgent memo: revised.

the procedure suggested in Chapter 4, they listed the facts (Figure 9.8), established a priority message, and re-wrote the memo selecting a problem/solution format (Figure 9.9).

    The second version fulfills the reader's request for information because it gives him what he really wants: the decision that must be made and the facts he needs to make it. Your writing is serving both you and your company well, when what you write makes your immediate supervisor's job easier.

Meeting Reports

Another type of solicited information memo is a meeting report. If all that is required is a record of what took place and who said what to whom in what order, then you can record the meeting and have a transcription made. The official minutes can be an edited version of the transcription.

LAKE ELECTRIC

Lamp and Electronic Parts Products Departments

TO: Jane Blake

SUBJECT: Purchasing take-up and accumulator equipment for CL-635

I have received a verbal quote from Ember-Mellon (EM) on take up and accumulator equipment.

Take Up	—$10,000.00—	Request 2	$20,000.00
Accumulator—	4,800.00—		4,800.00
		TOTAL	$24,800.00

I have requested EM by phone several times to supply me with formal quotes and prints on take-up controls and accumulator. I have explained that the prints should be reviewed by W. G. McCord, T. Gold, and W. C. Tilton of AME before purchasing equipment and that we need delivery by year end.

It doesn't appear that EM is very interested from their response, and their quote, $24,000, overspends the money on CL-635 by $8,400. However, their take-up equipment does have demonstrated reliability.

Attached please find a letter to EM explaining our position and requesting them to reduce quote.

I have another quote from Intel Machine Company (IMC) for double take-up unit @8,500 but not tape accumulator. I plan to review the status with W. G. McCord. In order to get this equipment by year end, I'll have to write the purchase order by the end of September; however, I want AME's blessing on any equipment that we purchase.

Also for your information, the original proposal for take-up and accumulator was to purchase take-ups and AME design accumulator. In mid-July, EM gave me a ballpark figure of $15,000 for building both take-up and accumulator. W. G. McCord and I both agreed that we couldn't build the equipment for $15,000.

Ralph Cole

cc: W. G. McCord,
    D. C. Forrest,
    W. C. Tilton
    T. Gold

Figure 9.7   Information memo: first draft.

The Facts

1. verbal quote from EM on take-up and accumulator equipment

	TU 10,000 per-	required 2	$20,000
	AC 4,800		4,800
		Total	$24,800

2. I have requested EM several times by phone to supply formal quote and prints for TU and AC because

    a. the prints should be reviewed by W. G. McCord and T. Gold of AME before equipment is purchased

    b. we need delivery by year-end.

3. E.M. doesn't seem very interested because

    a. they haven't provided the quotes in writing

    b. their verbal quote overspends the budget on CL-635 by $8400.

4. I've been waiting for EM this long only because of the proven reliability of their TU equipment.

5. I'm enclosing a copy of a letter which explains our situation to EM and requests a reduction in their quote.

6. Intel Machine Company has quoted me $8500 on a double TU without tape accumulator.

7. In order to get this equipment by year-end, I must write up the purchase order by September 30th. However, I will check its status with Bill because I want AME's approval on equipment purchased.

8. The original proposal was for AME to build the accumulator itself.

PRIORITY MESSAGE

A decision agreeable to both Product Engineering and Manufacturing Engineering must be made regarding the purchase of take up and accumulator equipment for CL-635 in time for the purchase order to be written by the end of September. Two bids have been made, one by Ember-Mellon and one by Intel Machine Company. EM equipment has a more proven reputation, but they have been lax in providing formal quotes and prints. IMC can provide the take-up but not the accumulator, whereas EM can provide both and can build the accumulator for less than it would cost AME to build one themselves.

RECOMMENDATION

I suggest we wait x days for EM to respond to my letter. If their response is not satisfactory, we will still be able to accept IMC's bid in time for end-of-year delivery.

Figure 9.8   Information memo: list of facts.

SUBJECT: Ordering of Take-Up and Accumulator Equipment for
             CL-635 in Time for End-Of-Year Delivery

Problem

A decision agreeable to both Product Engineering and Manufacturing Engineering needs to be made regarding the purchase of take-up and accumulator
equipment for CL-635. Bids have been made by both Ember-Mellon (EM) and Intel
Machine Company (IMC), and the purchase order must be written by the end of
September to ensure delivery by year-end.

Comparison of Choices

EM equipment has the more proven reliability, but they have been lax in providing formal quotes and prints, and their verbal estimate of $20,000 for two
take-ups and $4,800 for an accumulator (total $24,800) exceeds our project's budget by $8,400. IMC has provided a written bid for a double take-up for $8,500
without accumulator, whereas EM claims verbally that they can build the accumulator for less than it would cost AME to build in-house, as was originally
intended. Both W. G. McCord and I agree that we could not build the equipment
for their $15,000 estimate.

Recommended Solution

I suggest that we wait one business week for EM to respond to my final request
(see enclosed letter) to provide a formal bid and prints and to lower their initial
estimate. If their response is not satisfactory, we will still be able to decide
whether or not we wish to go with IMC in time for end-of-year delivery. Of course,
I will review all bids and prints received with the parties concerned in AME and
make certain I have their approval before I authorize purchase.

Figure 9.9   Information memo: revised.

If your purpose is to recall what took place, your minutes are a chronological
narrative—first this happened, then that . . . and so on. Figure 9.10 reproduces
the minutes of a meeting called to discuss the Ion Engine Flameout problem at
the Cuyahoga Lamp Plant of GE. The writer describes the topics as they come
up for discussion, numbering each, and the significance of the meeting is
revealed to the reader just as it was revealed to the participants—at the very
end:

> Hence the solution to the "ion engine flameout" problem shall be to utilize the
> portable flushing trays with the back-up assistance of either baking or light-up of
> A.T. after dose.

Again, if the purpose of your writing is to reconstruct what took place at the
meeting, this format may be appropriate. More often than not, however, to anyone requesting a report (memo) on the "ion engine flameout" meeting, the
order in which events took place or the way in which the participants accom-

ION ENGINE FLAMEOUT—MEETING MINUTES[1]
4-18-79

1. Cost estimate of the Ion Engine Flameout problem at Plant X is approximately $120,000. This estimate is based on the direct material and labor cost for all ion engine types except model #36 which is generally unaffected by flameouts. The breakdown in annualized cost is as follows:

Loss of .05% of all pellets made	$ 40,195
Loss of finished engines	75,763
Labor to restart engines	3,715
	$119,673

   Staff engineers are working with personnel at Plant Y to estimate their engine flameout cost. Because the severity of the problem seems to be less at Plant Y, the total cost between the two plants is at least $150–200K.

2. Batch baking of filled cesium pellets (350°C @ 10 min) was discussed as an effective method of reducing flameouts. However, the following problems prevent its use as our primary solution; they are:

   a. Should pellets be held for 24 hours between the filling and baking operation to eliminate leakers? If so what would be the in process inventory cost?

   b. Can oven trays be designed to contain the cesium in case of a pellet seal failure inside the baking oven?

   c. How will clean-up of contaminated pellets and the baking oven fixtures be handled? What will be the labor cost for extra handling?

   Therefore in view of the above problems we will consider using the baking operation as a back-up system only. Thus if used in conjunction with the portable flush trays the resulting quantity will be minimal, and can be more easily and effectively handled. Another back-up operation which must be studied is igniting pellets after fill. Both ideas are being investigated.

3. The dry box concept has been determined unfeasible at this time due to the high cost, approximately—$172K.

4. A new system of using portable flush trays was devised to prevent pellet exposure to air. This concept involves removing fixtures from the plasma welding and immediately placing them into a flushed holding container. Pellets would remain in the container until cooled then placed in portable flushing trays. The trays can be transferred from station to station and flushed at each one to prevent exposure to moisture (air).

   This concept is probably the most economically feasible and most effective method of dealing with the engine flameout problem. A cost analysis and an implementation study will be performed by the group. A back-up system of baking or igniting of pellets after fill may be used. Investigations are under way.

5. Cost estimates on metallic end caps proved to be quite expensive. Direct material and labor cost of $149K were estimated. In addition the effectiveness rate of plasma welding is only 70%. Therefore further consideration of its use has ceased.

6. The additional cost of handling and material for the polyethylene end caps coupled with their ineffectiveness concludes their use in conjunction with other solutions. The investigation of maintaining pellets at elevated temperatures from welding to filling has also been concluded because such maintenance cannot be implemented.

7. The cold chest concept is being studied for its use in conjunction with the portable flushing trays. There is some question as to its usefulness.

Hence the solution to the engine flameout problem will be to utilize the portable flushing trays with the back-up assistance of either baking or igniting of pellets after fill. The study is underway and will be presented at the next ion engine meeting.

[1] *Appropriate changes in names and technical terms have been made to protect the company's privacy.*

Figure 9.10   Meeting minutes.

plished their goal will not be of primary concern. Both to those involved who did and those who did not attend, the key questions needing to be answered are:

1. How was the problem defined?
2. What possible solutions were considered?
3. Which was chosen and why?
4. What direct effect, if any, does the decision have on what I do?

Using the first version as a guide, the author re-wrote his minutes with the conclusion first (Figure 9.11). He shortened it considerably by relegating some of the very technical details to attached figures. In this way, those needing them would have them and those not concerned would not have to "wade through" the sauce to get to the meat.

The same principle applies to the information memo known as a trip report. If to justify expenses your office needs an account of where you went and what you did, a narrative may suffice. Even better, your company may have a trip report form which enables you to fill in the blanks. But if the requestor's priority is to find out what the purpose of your trip was and what you accomplished, then you need to write a memo regarding your trip in which you analyze rather than just recount.

Requesting Information

If you are the one writing to ask for information, think in terms of why the person or group would want to give you the information and why they would

ALTERNATE VERSION

Ion Engine Flameout—Meeting Minutes
4-18-79

Summary

The purpose of the meeting was to discuss progress in the investigation of seven possible solutions to the ion engine flameout problem. Since all other alternatives have proved ineffective or too expensive, it was decided to utilize the portable flushing trays with the back-up assistance of either batch baking, igniting pellets after fill, or the cold chest concept. The results of more detailed study will be presented at the next ion engine meeting on Thursday, May 23, 1979.

Background

Cost estimate of the engine flameout problem at Plant X: $120,000.
— based on the direct material and labor cost for all ion engines (except model #36 which appears unaffected).
—for breakdown of annualized cost see attached Table 1.

Cost estimate of engine flameouts at Plant Y: c. $80,000.

Total cost to both plants: $150–200K.

Analysis

Of the seven possible solutions considered, two—1. the dry box concept and 2. metallic end caps for plasma welding—have been eliminated due to excessive cost. Two others—1. maintenance of pellets at elevated temperatures from welding to filling, and 2. batch baking of filled pellets—were also eliminated due to technical difficulties restricting implementation.

The most economical and technically feasible solution is a new system of using portable flush trays devised to prevent pellets' exposure to air in combination with one of three possible back-ups: 1. batch baking, 2. igniting of pellets after fill, or 3. the cold chest concept.

Notes: a. For a detailed explanation of the portable flushing process see Figure 1.

b. For a detailed comparison chart of possible solutions considered, see Figure 2.

Figure 9.11   Meeting minutes: alternate version.

not. Anticipate the needs of your readers in your request letter by:

1. making sure you give them all the information they need to know to supply you with what you need to know
2. making it as simple as possible for them to fulfill your request (i.e., providing return postage or a form to fill in; giving a reasonable deadline or suggestions for where they might locate what you need, etc.).

Taking the "You" Attitude

Pay attention to the attitude or tone of your request. For example, instead of "I need your metals report to complete my metals report," try "I'm sure we both desire to compete successfully with larger companies in our area and might each benefit from pooling our resources. It would assist me in preparing my metals report to be acquainted with the direction your own firm is taking. I would, of course, be willing to offer you a copy of our report in return. Naturally, it is understood that any confidential information may be omitted from the copy you send us."

Be considerate of the reader by providing in your request the precise information the reader needs to reply to you, effectively.

Here is how *not* to request and give information:

The Henning Corporation
Sales Department
Timberville, Georgia

Dear Sir:

    I want to build a room divider. Do you have what I need?

                  Sincerely,

                  Joseph Short

Mr. Joseph Short
3 Rocky Road
Cliff, Idaho

Dear Mr. Short:

    Yes, we do!

                  Sincerely,

                  Bill Blaine
                  Customer Services Manager
                  The Henning Corporation
                  Timberville, Georgia

Instead, be thorough and specific. For example:

Sales Department
The Henning Corporation
Timberville, Georgia

Dear Sirs:

   I want to fix a room divider using perforated hardboard panels, cutting and grooving my own framing. Does your firm carry the tension parts that go at the top of the framing?

   Also, if the divider needs to be eight feet long, what size lumber should be used for the framing? Will the panels sag if left the four foot width? They are to be six feet high with open space above and below. I would appreciate an economical solution.

   Thank you for your time and interest.

                    Sincerely,

                    Joseph Short
                    3 Rocky Road
                    Cliff, Idaho

Mr. Joseph Short
3 Rocky Road
Cliff, Idaho

Dear Mr. Short:

   Thank you for your inquiry about room dividers.
   Our company does stock what we call "timber toppers." The #52 is hardwood and fits a closet pole 1-⅜ inches in diameter. The #50 is steel and fits a standard 2 x 3. These spring-loaded devices make it easy for you to design your own tension pole-type construction.
   If your local hardware store does not stock these items, you can order them directly from us at $4.25 each.
   As the enclosed diagram shows, you might build the divider you want using the #50 timber toppers. Perforated hardboard 48 inches wide is apt to bow without a supporting frame. We recommend a light 1 x 2 frame that will bolt between three 2 x 3's as shown.
   Affix the hardboards to the frames and trim out with something decorative if

you like. Use t-nuts in the 2 X 3's to receive the bolts, and you'll have a neat, demountable, reusable structure.

 I trust this information helps, and have enclosed our catalogue to acquaint you with other products you may require.

                              Sincerely,

                              Bill Blaine
                              Customer Services Manager

 The more knowledgeable and considerate you show yourself to be, the more likely you are to get the full cooperation of the reader from whom you are requesting information.

 The request letter in Figure 9.12, for example, might have been written as a complaint letter. Quality Control representatives had been turning up randomly and making unexpected requests that confused the operators and interrupted their accustomed schedule. Since the operators' foreman had not been apprised of Quality Control's objectives, he could not prepare his workers to cooperate with the Quality Control engineers. Rather than complain about this communication failure, the foreman chose to request the information he needed to

Figure 9.12   Request memo.

TO: Don Winters
         Director, Quality Control

FROM: Bill Evans
         Foreman, Plant #3

RE: Procedures for Exhaust Machine Quality Checks

Since it is in the best interests of both our divisions to keep the exhaust machine functioning at peak efficiency, I am enlisting your aid in responding to recent questions from the machine operators regarding Quality Control's procedures for product checks.

The operators are confused by repeated, supposedly unscheduled requests for lamp samples. If they had Quality Control's schedule, the operators could cooperate by providing your engineers with the samples and information they need to do their job with minimum interruption of their own responsibilities.

Please provide a lamp check schedule, and let me know if there is any other way that we can facilitate each other's operations.

solve the problem. Since the Quality Control Director did not feel criticized by this approach and was able to see the good reasons behind the request, he provided the information (Figure 9.13).

Given this information, the operators feel more like part of a team effort and less imposed on. If any of the terms outlined are unreasonable from their perspective (if they feel it is not practical to get the lamps to a quality control engineer in one half an hour or less for instance) representatives of each group can get together and work out the problems.

Perhaps, the most important point to learn about giving or requesting information is how to know when you are in a situation in which writing these kinds of memos will improve employee relations.

COMPLAINING AND RESPONDING TO COMPLAINTS

Very often what started off as a lack of sufficient information becomes a need to complain. Whether you are writing or responding to a complaint to or from a customer, a supplier or a co-worker, you are still writing an information letter. If you are the complainer, you must give the reader as much information as possible:

1. Describe the product or situation (including serial and model numbers, dates, costs, names, title, addresses, etc.).
2. Describe what you want.
3. Be clear about what you will or will not accept as a solution to the problem.
4. Give the reader a deadline.

Figure 9.13   Reply to request memo.

TO: Bill Evans	Date: February 21, 1980
FROM: Don Winters	
RE: LAMPS REQUIRED FOR EXHAUST MACHINE QUALITY CHECKS	

In order to ensure maximum operating efficiency, the quality checker needs the exhaust machine operator to provide lamps according to the following schedule:

Fill Pressure: 4 lamps per machine at each start-up or change-over with a minimum of eight (8) per shift.

Head Check: 1 lamp from each head at each start-up or change-over.

NOTE: In order for inspections to be accurate, lamps must be provided for the quality checker within one half hour of the quality checker's request. If a substandard product is detected, repeats will be required.

With this information, the reader can determine how to respond to your complaint quickly and efficiently.

If you are the recipient of a complaint, you need to provide your reader(s) with the following information when you respond:

1. Whether or not your understanding of the circumstances agrees with theirs and, if not, how it differs.

2. Whether or not you can fulfill their request, when and how, and, if not, why not.

3. What, if any, alternative solutions you have to offer.

4. What, if any, restrictions apply regarding their choice of what to do next (i.e., schedules, deadlines, legal requirements, etc.).

Receiving a Complaint

The purchaser of a vacuum cleaner writes to the manufacturer because she thinks she has a lemon (Figure 9.14). In responding to this complaint letter, the recipient's first step is to analyze its content carefully.

* Consider the *facts* and the *attitude*.
* Decide what the writer wants; and whether you want and are able to supply it. Only then will you know if you are writing a letter of compliance or a letter of refusal. Saying "yes" requires a different structure and tone than saying "no."

In the *first paragraph* of the complaint letter, the customer, Mrs. Brown, identifies herself as a long-time patron (i.e., someone valuable to you). She also identifies the model of her machine, how long she has owned it, that she has a two-year warranty, and that she has had to take the cleaner to the repair shop with the same problem three times in the past nine months. It may occur to you that the customer has chosen an inept repairman. But she anticipates this response in the opening of the second paragraph by making clear that he is running one of your authorized repair shops.

The *second paragraph* tells what is wrong with the machine, suggesting rather strongly that a new machine should not have these problems.

The *third paragraph* is the customer's argument that she is not responsible for the machine's malfunctioning because in the company's advertising and instructions on how to use the vacuum, they did not indicate the need for special coddling.

The *fourth paragraph* contains a thinly veiled threat that you may lose the patronage of this long-time customer.

Finally, in the *closing paragraph* the customer makes her request in the form of three choices:

1. Fix the machine once and for all.

2. Give me a new machine.

3. Give me my money back.

March 23, 1979

Director of Customer Service
E-Z Vacuum Cleaner Company
Pine Building
Little Rock, Arkansas

Dear Sir:

I have owned E-Z vacuum cleaners for over twenty years. In September of this year I purchased your new upright model with a full two-year warranty. In the past nine months, I have had to take the machine to the repair shop three times, in each case because it was getting "stuffed up".

The repairman I went to, John Smith, runs an authorized service center for your vacuum cleaners. The first time I came in, I was told that I wasn't cleaning the filter properly. The second time I was told that its "normal use" clause in my warranty didn't apply because I had vacuumed up some paper clips. The third time, I was told that I wasn't allowing the machine to "breathe" properly, and should be sure to hold it tilted at one end for at least 90 seconds after each use.

To be blunt, I am a little tired of being accused of machine abuse. Furthermore, since it does not say anywhere in the instructions for using the vacuum cleaner that it is necessary to let it "breathe" after each use, I do not feel you can hold me responsible for misusing it because I have not let it "breathe".

If the new upright is really so temperamental, your advertising description and user manual should make that clear. I for one would not have brought this model, had I known how difficult it was to avoid clogging it.

Please either honor your warranty, give me a new machine; or give me my money back.

Sincerely,

Mrs. Morton J. Brown

Figure 9.14 Complaint letter.

Which one will you choose? Can any of the choices be eliminated immediately? Probably, you will eliminate the third one. Your company most likely has a policy about how long a customer can own a product and still be allowed to return it for a full refund. Nine months is probably too long for her to say, "Take it back; it's yours!"

So the question is, can Mrs. Brown's machine be fixed? Is it a "lemon" that should be replaced, or is there an alternative that Mrs. Brown has not mentioned?

To answer these questions you have to do some research and collect some additional facts. First of all, you will probably contact her serviceman, Mr. Smith, to get his version of the problem. What he might tell you would shape your reply to Mrs. Brown.

Possibility 1

Mr. Yates, Customer Service representative of the E-Z Vacuum Cleaner Company, calls Mr. Smith:

MR. YATES: I have an angry letter here from a customer who claims that both our vacuum cleaner and your shop have been giving her a hard time. I wanted to get the facts from you before I answered her complaint.

MR. SMITH: What's her name?

MR. YATES: Mrs. Brown.

MR. SMITH: Oh, her! That woman has tried to pick up everything but her kitchen sink with her new vacuum cleaner. Each time I patiently unclog it, and each time she zooms off with it again and re-clogs it. I think I must have reread the instructions to her out loud at least ten times. What she needs is to trade her vacuum cleaner in for a derrick.

MR. YATES: I see. Then she really is abusing the machine. Well, maybe a letter from us confirming your diagnosis will help.

Possibility 2

Mr. Yates calls Mr. Smith:

MR. YATES: I have an angry letter here from a customer who claims that both our vacuum cleaner and your shop have been giving her a hard time.

MR. SMITH: What's her name?

MR. YATES: Mrs. Brown.

MR. SMITH: Yes, I remember her. And to tell you the truth, Mr. Yates, she's not the only customer having problems with model 662B. In fact, if you hadn't called me, I probably would have called you. At first, I did blame it on customer abuse, but frankly I'm beginning to think that either a batch of lemons was sold in our area or else there's a flaw in the basic design of this cleaner.

MR. YATES: Thank you for being honest, Mr. Smith. I'll check into the problem further and get back to you.

There are many possible variations on the two telephone conversations just presented. Those in the first category would provide verification that the customer is responsible for the problems with the product. Those in the second category would provide evidence that the product, and the manufacturer, might be responsible for the malfunctioning.

If your phone inquiry produced information falling into the second category, indicating possible manufacturer liability, you would want to investigate further. You would probably requisition available files on the rate and type of complaints regarding model 662B throughout Mr. Smith's service area and throughout the areas in which you sell. You would also want to look at figures on frequency of repairs done during the tenure of the warranty on that model. Only then could you determine if Mrs. Brown's problem is typical or atypical. Because such investigations take time, you will want to send Mrs. Brown an interum note, informing her of what is being done and approximately when she will receive an answer.

Whether or not you conclude that your company or Mrs. Brown is responsible for the frequent malfunction of the product, you will want to have time to assemble all the facts *before* writing your final determination.

As in all your writing tasks, you are now going to combine what you know about the problem with what you know about the audience in order to formulate an effective response. Depending on the circumstances, your reply to Mrs. Brown will either be a "good news" or a "bad news" letter.

Writing a "Good News" Letter

In a "good news" letter, you begin with the good news. In this case, you are going to tell Mrs. Brown something she wants to hear. You are going to offer her a new machine, a free repair of the one she has, or something else that will please her. Use the general to specific structure of writing: a general statement of what you will do to satisfy Mrs. Brown followed by detailed explanations and instructions, and concluded with a pleasant, "glad to have served you" close.

What is meant by opening with a general statement of the good news? Would this do?

Dear Mrs. Brown:

   The E-Z Vacuum Cleaner Company thanks you for your recent inquiry regarding the malfunction of your 662B vacuum cleaner.
   We want you to know that we will do everything possible to make you happy.

                    Sincerely,

                    Mr. Yates
                    Customer Service Representative

What does "everything possible" mean? and from whose point of view? Is the writer taking the "You" or the "Me" attitude? Such an approach is vague rather than general and can do more harm than good. Figure 9.15 presents a more effective letter.

Under these circumstances you will have a second letter to write to the serviceman, Mr. Smith. You will want to keep his good will; to fully explain the problem in the more technical terms that a mechanic can understand; and to make sure that he comprehends how to handle the customers who will be bringing the faulty vacuum cleaners back to him.

Your letter to Mr. Smith is not entirely "good news" or "bad news," but a delicate combination of both.

Once again, you adopt the "You" attitude. There he is, busy at work on a vacuum cleaner that some customer has used to sweep up thumb tacks. Surrounding him are other uprights, tank tops, and dialamatics waiting to be fixed. In the background, the phone rings impatiently. At this moment, the mailman arrives with your letter (Figure 9.16).

Writing a "Bad News" Letter

In addition to the "good news" and "mixed news" format, both of which begin with a general statement of the good news and go on to the specific details, there is the refusal or "bad news" letter. The "bad news" letter begins with the specific details that gradually prepare the reader to receive and accept the statement of rejection or refusal. It is one of the few exceptions to the thesis of Chapter 4—that starting with the conclusion is the clearer way to write.

When you refuse:

- present facts and reasons *before* refusing,
- offer a compromise or alternative, where possible,
- close on a positive note.

There is no deception or mystery in this type of letter writing, however, when it is done properly and responsibly. The purpose of this format is *not* to "put off" the news or keep the reader guessing. Most readers sense the refusal implied in the lack of immediate acceptance. The idea is to get them immediately involved in your reasons rather than in their emotional reaction. While they are engaged in interpreting and reacting to facts and opinions, your readers have no time to be angry or disappointed. By the time they get the actual statement of refusal, they may even agree that you had legitimate reasons and that they might have done the same in your position. Explain first, and say "no" later.

Figure 9.17 is a sample "bad news" letter that might have been written to Mrs. Brown.

Complaints to, about, or by You

Many of you may never have direct contact with your firm's customers or clients and may never need to respond directly to a customer complaint. But regardless

March 29, 1986

Mrs. Morton J. Brown
24 Northwood Drive
Little Rock, Arkansas

Dear Mrs. Brown:

Thank you for your letter of March 23, 1986 in which you explain the difficulty you have been experiencing with your E-Z vacuum cleaner #662B. You can rest assured that all the necessary repairs will be made on your machine immediately and at no cost to you.

You, along with many of our faithful customers in sales region #3 were unfortunately victims of a production error at our Eastern plant. You are absolutely correct in stating that the instructions for operating your machine should include any special directions necessary to avoid clogging. However, such instructions were not included because the properly built 662B machine doesn't clog in normal use.

Unfortunately, several hundred machines, including the one you purchased, were built with the agitators slightly off balance. It is the agitator propelling the brushes improperly that is making your machine "choke".

Naturally, we are recalling the "lemons" at our expense in order to replace the agitators.

If you would be so kind as to take your cleaner to Mr. Smith just one more time, he will replace the faulty parts as soon as we have shipped the good ones to him, and he will notify you when your cleaner is in proper working order.

Please accept our sincere apologies for this inconvenience and feel free to contact me personally in the unlikely event that you have any further problems.

Sincerely,

Mr. William Yates
Customer Service

Figure 9.15   Good news letter.

March 29, 1979

Mr. Lloyd Smith
Smith's Repair Shop
12 Newbold Street
Little Rock, Arkansas

Dear Mr. Smith:

You will be pleased to hear that we have discovered what needs to be done to end the frequent complaints you've been getting from Mrs. Brown and other recent purchasers of our 662B upright.

It appears that in one shipment of cleaners, the #2¼ agitator was installed instead of the required #2, causing the brushes to rotate at a slight angle. Every time the agitator rotates on these machines, it loosens the beater bars and twists the belt. This in turn overloads the motor and traps dirt in the machines. No wonder they can't breathe.

We are recalling the affected machines and feel sure that you will want to cooperate with us in maintaining our mutual customers' good will. We have written to the customers involved, instructing them to bring the machines to you for replacement of the faulty part. We'll ship the replacement parts to you at our expense.

We know you'll be especially courteous to the affected customers and promptly repair and return their machines. Send us the bill, of course.

Naturally, this extra work cuts into your already busy schedule. We sincerely apologize. We trust you will agree that it is better to make the best of a bad situation then to leave several good customers antagonized and dissatisfied.

You have our heartfelt apologies for the inconvenience and our gratitude for your cooperation in maintaining the regular clientele on which we both rely.

Sincerely,

Mr. William Yates
Customer Service
E-Z Vacuum Cleaner Co.

Figure 9.16 Mixed news letter.

March 29, 1979

Mrs. Morton J. Brown
24 Northwood Drive
Little Rock, Arkansas

Dear Mrs. Brown:

Naturally, having been a faithful user of E-Z vacuum cleaners for so many years, you are upset to suddenly experience a difficulty with one of our products. On top of that, it must be terribly frustrating to bring your machine in time and again for repair, only to be back where you started. And although I'm sure Mr. Smith was only trying to help, it is, of course, difficult for a mature homemaker to be informed that she doesn't know how to use a vacuum cleaner properly. You have our sympathy and our apologies for the inconvenience you have suffered.

Anxious to get to the source of the problem, we immediately scanned our repair and complaint records to determine the frequency of similar problems with the 662B. Actually, based on our projected averages, the 662B tends to have an above-average low-repair frequency of once every three years. Of course, since the model has only been on the market for 1½ years, this is an estimate rather than an actual rate. Nevertheless, our past repair rate estimates have been 98.9% correct.

We also noted that there is no significant correlation in the types of problems experienced by users of the latest upright model. This observation leads us to believe that Mr. Smith, who as you know, has an excellent twenty-year service record, may be partly correct in his analysis of the problem. Perhaps the solution would be to help you become better acquainted with the operation of your new machine.

For example, Mr. Smith tells me that your last E-Z vacuum cleaner was a cannister type. The cannister cleaner relies on direct suction to pick up dirt, while the upright combines two operations for cleaning; an agitator brush for dislodging dirt from carpeting and suction to pull dirt into the bag. A switch from one model to the other takes some getting used to, understandably.

Since it does appear, then, that the deficient operation of your vacuum cleaner is neither our fault nor Mr. Smith's, we cannot provide a solution by any of the means you suggest. We can offer you, however, one free check-up for your machine by Mr. Smith.

If you will be so kind as to bring the vacuum cleaner back to him one more time, we will authorize a thorough inspection, part by part, and he will explain the operation and maintenance of each part to your satisfaction.

Thank you for your interest and your patronage.

Sincerely,

Mr. William Yates
Customer Service
E-Z Vacuum Cleaner Co.

Figure 9.17 Bad news letter.

of what position you hold in your company, you perform tasks. You do something, and anyone who does something lays him/herself open to the criticism and complaints of others about how it is done. As with the vacuum-cleaner example, sometimes these complaints are legitimate; sometimes, they are not. Part of doing your job is knowing how to respond in writing to both justified and unjustified dissatisfaction.

By the same token, because we all must work with others and depend on whether other people keep up their end, we will have to complain occasionally to or about other people. Following is a list of common situations in which you might be complained to or about or might need to complain to or about someone else in the course of your daily responsibilities. By adapting the same procedures outlined in the vacuum-cleaner example, you can handle each of these predicaments effectively:

1. A subordinate complains that you are assigning too much work, not supplying him/her with adequate backup, or not providing enough advance notice for the job to be done properly.
2. A person or group that you supervise complains about unsafe equipment or unhealthy working conditions.
3. One or more members of your staff complains to you about the allegedly poor work habits or interference of another member of your staff.
4. A supplier complains that you are not meeting your part of the contract. For example, he cannot supply custom-designed parts on time, if your design engineers are not on schedule with their specs and diagrams.
5. The office accountant complains that she cannot provide an accurate audit because your records and receipts are insufficient.

In each case, either you really have been remiss or the complaint is based on inadequate information, or different interpretations of the same information. If you are in the wrong and there is something you can do about it, your reply should admit responsibility and state specifically what you will do to rectify the situation and when. If you check out the facts and believe you are not at fault or that there is nothing you can do to change the condition complained of, explain your reasons to the complainer and discuss what "we" can do about "our" problems. Suggest a compromise or alternate solution whenever possible. If you cannot reduce the noise in an employee's work area, maybe you can find a quiet place elsewhere in the building that he can go to when he needs to write a report or think out a project. Maybe you can arrange for him to work at home one day a week.

Complaints You Might Make

1. You are a computer program analyst responsible for setting up a new information system for the Advanced Mechanical Engineering (AME) Department of your firm. Before you can evaluate vendor packages for the project, you need to compile and analyze data on the AME information flow in each group within the department. But the managers of each group refuse to allot

time for being interviewed, and they are the only ones who can provide the input you require to do your job. Your boss is complaining that you will not meet your deadline, a fact that your latest progress report confirms. You will have to complain about the lack of cooperation either to your boss or to the AME managers directly.

2. Your manager has purchased two large pieces of equipment from a German manufacturer. Your group will utilize this equipment which will replace an American model. The foreign manufacturer offers training in the use and maintenance of its products at its offices in Hamburg. But your manager has not alloted money in the project budget for training and refuses to send two of your engineers overseas. The time it will take for the staff to figure out for themselves the operations and capacities of the equipment and the errors made along the way will be costly. You need to complain.

3. You have been working on a project team that you really like in an area that enables you to develop the expertise you already have and learn some new skills as well. Halfway through the project, you are asked to take a new assignment so that another metallurgist who is not working out well on some other team can give yours a try. You feel it would be unfair to accommodate the other employee at your expense.

4. Quality Control is called in to determine why a fluorescent shell washer is not functioning properly. It turns out that the new cleaning agent purchased from an outside supplier contains an adhesive substance which causes the fluorescent light shells to stick together and jam the conveyor belt. You contact the supplier's representative by phone, pointing out that the need for a non-stick fluid was specified in your contract. He informs you that he needs the complaint in writing before he can take action.

Your first reaction to being in a situation that calls for a complaint will probably be annoyance, anger, frustration, or resentment. These are normal first reactions. However, expressing them in your formal letter of complaint is the least effective way to get the action that you want from your reader.

Think about how *you* react if someone tells you, "This is all your fault. Look at all the trouble you've caused. I'm furious at you." Instead of wanting to help them, you want to fight back, to defend your position, or just to outdo them in name-calling. These are not the reactions you want from your readers in a professional situation. The computer program analyst in example 1 does not want the AME engineers to fight back; she wants them to cooperate on her project. The metallurgical engineer in example 3 does not want his managers to feel criticized or threatened; he wants them to understand and cooperate in his career goals. He does not want to question the worth of the other person involved but to focus on his worth.

If you find your emotions regarding a complaint are getting the better of your judgment, give way to your emotions privately by writing them out. Write the nastiest letter you can imagine and read it through as many times as it takes you to vent your feelings. Then destroy this letter and write the one that you will actually send. Having cleared the air, you can analyze the situation more objec-

September 15, 1986

Mr. George Burke
Ember-Mellon
33 Channing Rd.
Phoenix, Arizona

Dear George:

I have received your verbal quote of $10,000 per take-up and $4,800 for
accumulator, per phone call with Herb. The quote is higher than I had
anticipated. I only have approved money of up to $16,000 for take up and
accumulator.

I would really like to use your equipment because of its past reliability. If
there is any possible way of reducing your quote, give me a call and send me a
formula quote with prints. I have to review the prints and quote with
Cleveland Engineering. Also, we need the equipment by the end of 1986 as
money is approved to spend in 1986.

I have another quote for take up equipment and will have to issue a purchase
order to them by the end of September unless I hear from you.

Thank you for your time. I would prefer to issue the purchase order to
Ember-Mellon.

Sincerely,

Ralph Cole

Figure 9.18 "You" attitude complaint letter.

tively and write a communication that will explain your reasons for complaining
in a way that makes your reader willing to consider and even give you what you
want.

More often than not, you will discover that your complaint letter is not really
a complaint at all. It may take the form of an information letter which, through
the strategy it uses to present the facts, persuades the reader to rectify the problem without a complaint ever actually being made.

You recall, for example, the information memo that Ralph Cole sent to his
boss regarding bids for take-up and accumulator equipment (Figure 9.7).
Attached to it is a copy of the letter he sent to the company he really wanted to
order from, but could not because they were delaying submitting a bid in writing. Cole could have written a letter complaining of the delay. Instead, he did
his complaining at lunch with a friend who was working on the same project and

178 Brookline Avenue
Boston, MA 02118
April 21, 1987

Mr. Stanley Cutter
1745 East 116th Place #23
Cleveland, OH 44106

Dear Mr. Cutter:

In response to your letter of April 19, I must reject your proposal for use of
the hospital computer for the following reasons:

1. State and Federal laws require that patient records be kept confidential. The
   policy of Beth Israel Hospital is to allow only licensed physicians access to
   patient records and then only under strict confidentiality regulations.
2. Summer research grants are allocated only to students who have completed
   at least one year at Harvard Medical School. Applications for these grants
   must be filed by the March 15 preceding the summer that they will be used.

There are many research opportunities which do not require access to patient
records and that will be available this fall at Harvard Medical School and its
affiliated hospitals. Most can be taken for credit, and many have stipends
available. I believe that you will find many of them interesting and
worthwhile.

I look forward to welcoming you to the Harvard Medical School community
this fall.

Sincerely yours,

B. G. Schott, M.D.
Director of Research
Beth Israel Hospital

Figure 9.19   Refusal of request.

could share his annoyance. To the tardy bidder, however, he used the strategy
of pointing out reasons from the reader's point of view that the written bid
should be sent out immediately and should be lower than the verbal bid (Figure
9.18).

PERSUADING OR RESPONDING TO PERSUASION

As you can see, all letters and memos and even straight narratives such as meeting minutes and trip reports are attempts to persuade. When you request and

impart or refuse to impart information, or when you complain or respond to a complaint, you are persuading your reader to see the facts and the situation the way you see them.

Proper coordination of your facts with your reader's anticipated perspective produces a communication in which the information is persuasive in its own right. Presented cleverly, it convinces. Sometimes, however, the "human factor" in technical communications determines whether or not the information is persuasive. The letter in Figure 9.19, for instance, replies to a request by a student just admitted to medical school. He wants to use the teaching hospital's computer during the summer prior to his enrollment. The denial of the request combines reasonable explanations with a concern for the student's attitude toward the institution to which he is about to devote several years of his life. The aim of the letter is to persuade the student that the hospital has good reasons to deny *this* request, but will not necessarily deny all his requests.

The Danger of Ambiguity

Sometimes, to avoid being too aggressively persuasive or opinionated, we err in the opposite direction. The result is a communication which makes it difficult for the reader to make a decision because the writer's own position is ambiguous or inconsistent.

The letter in Figure 9.20 is adapted from one written by an international sales manager to an overseas distributor. When it was presented at a technical writing workshop, the group commented that the message was garbled. In accordance with the planning method explained in Chapter 4, the writer was asked to list her facts and determine her main message (Figure 9.21). With the greater understanding the listing technique gave her of the information and of the "human factor," she was able to revise the letter so that it respected her associate's freedom of choice while still using the facts persuasively (Figure 9.22).

Few people realize the extent to which their written interactions with others affect their ability to do their job. If you know when and how to request or give information, when and how to complain, and when and how to be persuasive, you can apply each of these skills whenever they are required. To sharpen your skills, keep files of effective letters and memos of different types—both those you have written and those written to you. Effective examples of your own writing help you recognize and develop your own style. Effective phrasing, paragraphing and even entire letters can be reused when appropriate, particularly given word processing facilities. Examples of your own correspondence will also help you remember what works well with each of the people you write to regularly.

In most professions, writing daily correspondence should be subordinate to your other job skills. But you can only subordinate a skill after you have become proficient at it.

Mr. John Doe
Manager, Acapulco Office
Taco Electronics
Acapulco, Mexico

Dear John,

This is in response to your letter of February 6, 1987 regarding the Eighth
World Computer Exhibition in Mexico City for September 10–12, 1987.

I am in agreement with your request to have all options and one of each plug-in
amplifier for the X2700 and Y750; however, at this time you know all of these
are not available. Also, there might be some question as to when these units
would be deliverable to customers with all of the options available.

It is my opinion that although I would like very much to have the X2700 and
Y750 for this computer exhibition, it would be in the best interest of both Taco
Electronics and XYZ Company *not* to participate with the X2700 and the Y750.
First of all, the exhibition is not for analog-type recorder products, and I think
it would be non-productive for XYZ Company.

We would like to recommend that when more of the options for the X2700 and
the Y750 are available, we conduct a private showing similar to what we did
last year with the biophysical scope monitor. We would spend about four to five
weeks in Mexico going from city to city for demonstration purposes. The reason
I say this is that all of the options are not and will not be available for the
initial deliveries of just the recorders.

If you feel that the demonstration of the X2700 and Y750 still would be
beneficial, I would do everything possible to have the basic recorder with as
many of the options as possible; however, not many of the characteristic
options for amplifiers and/or multiplexing, and/or digital inputs will be
available for September. If your decision is to go ahead with this, please
acknowledge by return telex because tentatively, I have initiated securing an
X2700 and Y750 for your September 1 deadline.

Very truly yours,

Joan Hall

Figure 9.20 Ambiguous response to request.

Dear John,

XYZ Company appreciates your eagerness to introduce our X2700 and Y750
recorders with all options and one of each plug-in amplifier at the Eighth World
Computer Exhibition in Mexico City on September 10–12, 1987.

We endorse your efforts to increase our distribution in Mexico, but have two
reservations about doing so by means of the Mexico City exhibition. First of all,
the analog-type recorder seems to be unrelated to the kind of instrumentation
that will be represented generally. Second, it now seems fairly certain that we
will not have all the options and the amplifiers ready for exhibiting in early
September. Even if an exhibition model can be set up, it is unlikely that we
could deliver units with all the options to our customers.

Given these circumstances, we feel it would be more productive to wait until
more of the options are available and then conduct a series of private showings
over a 4- to 5-week period, just as we did last year with the biophysical scope
monitor.

In response to your initial request, I did tentatively secure an X2700 and Y750
for your September 1 deadline. But given the alternative of private showings
later in the year with all the options ready to go, you might seriously consider
letting the Mexico City exhibition go by.

Please advise me of your decision by telex.

                                        Sincerely,

                                        Joan Hall

Figure 9.21   Revised version of ambiguous letter.

EXERCISES

Giving or Requesting Information

1. Write a memo to the members of your fraternity or sorority explaining a change in
   scheduling of chores for your residence; or write a memo to your roommate or fam-
   ily members explaining how household chores should be divided.
2. Write a memo to your co-workers (or club or team members) detailing a significant
   change in procedures.
3. Write a letter requesting an interview from a busy professional person for a term
   paper you are preparing. (Remember to take the "You" attitude.)

LETTER TO OVERSEAS SALES REPRESENTATIVE

What Are the Facts?

- The writer agrees that ideally the two models with all options should be exhibited in Mexico.
- The equipment with all the options is not available yet.
- Even if it were available by September 10th for exhibit purposes, units with all options might not be ready for delivery to customers.
- XYZ Company should probably not participate with the X2700 and the Y750 anyway, because the exhibition is not for analog-type recorders.
- XYZ should wait until more of the options are available and then enter the Mexican market via private showings. This method has worked before with the biophysical scope monitor.
- If John still wants to go ahead, despite these cautions, XYZ can get the equipment to him without some of the options.

What Do You Want to Tell the Audience? What Do You Want the Audience to Think or Do?

I want John to figure out for himself from the facts that his exhibition plan is not the way to go. But I want to convince him without dampening his enthusiasm for our products, his anticipation of good profits, and his self-confidence. I also want him to know that we appreciate his efforts and are not trying to preempt his judgment.

Figure 9.22   List of facts in memo.

4. Write a letter to a graduate school or adult education center requesting specific information to determine whether a program it offers would meet your career goals.

Complaining and Answering Complaints

5. Write a letter protesting a change in procedures or a lack of adequate procedures at your school or your job.
6. Write a letter complaining about a product you have purchased.
7. Write a reply to the letter you wrote for number 5.
8. Write a reply to the letter you wrote for number 6.

Persuading and Responding to Persuasion

9. Write a letter to the dean of a law school that admits full-time students only. Convince the dean to make an exception in your case and admit you as a part-time student.
10. Reverse your perspective, and write what you think the dean's response to your request might be.

11. Write a letter to a relative or friend to persuade him/her to invest $1,000 in a business scheme that you would like to try out (e.g., making your own jewelry or wooden toys, repairing bikes, painting houses).

12. After you have written the letter in question 11, reverse your perspective and write a reply to that letter, including any questions that the persuader has left unanswered.

13. Analyze the effectiveness of the progress memo that follows. Consider questions such as

    a. Is it a "good news" or "bad news" message?

    b. Is the structure appropriate to the intention?

    c. What changes, if any, would you make and why?

TO: The Boss
FROM: Me
DATE: 4/15/87
RE: MTPD Information Systems Study Progress Report

At this point, the MTPD Information Systems Study is progressing slower than I had anticipated. There are two major areas of the study underway at this point, both involved in the project planning and control phase of the study. These areas include:

1. Information collection and consolidation.
2. Evaluation of vendor packages for project planning and control.

It appears that progress in these areas should improve within the week. All other aspects of the study are inactive at this point.

The slowest area is the gathering of information from the department AME managers. It's been difficult to arrange meetings to discuss their department information flow. Fortunately, only two groups remain to be contacted.

Another slow area is the documentation of the AME information flow. The CAD system on which the documentation is being done has very slow response time when more than one user is working on the system. To alleviate this problem somewhat, one of the technicians has consented to work from 1–9 PM to utilize time on the CAD system better. This should speed up the documentation process.

With these two areas almost under control, the study should move ahead with much greater speed. The data from each AME department needs to be consolidated into one workable system. When all of the information needs are consolidated, I will proceed with the development of forms and the data flow needed to eventually automate the entire process.

The second major area of concern is choosing the system to do this work. The information I requested from outside vendors is beginning to come in. I am reviewing this information as it arrives and am trying to eliminate some of the systems according to the data I'm collecting from the AME groups. I am also constantly looking for other system packages.

Although progress has not gone as I anticipated, once the information is collected and documented, the rest should move smoothly. I see no problem in meeting the June 1 date for the final design of the project planning and control system.

CHAPTER 10

Writing Descriptions and Instructions

To write descriptions and instructions

- Know your material.
- Know your readers.
- Design the description or instruction and the accompanying graphic arts to suit a) the readers' level of expertise, and b) the purpose for which *they* will use the information.

Description of Objects

- Begin by establishing how the readers will use the description (why they need to know the object to do their job).
- Summarize the main features.
- Give the pertinent details accompanied by an easy-to-read, but thorough diagram.

Descriptions of Processes

To Readers Who Need to Understand How Something Works

- Include only those details pertaining to the reason the readers need to understand the process.
- Use analogies to things with which the readers are already familiar.

To Readers Who Need to Perform the Process

Provide

- an opening explanation of scope and goals,
- a complete explanation of objects and materials,
- a step-by-step description of what to do when,
- a separate discussion of special cases, exceptions, and emergency measures.

THE DANGERS OF POOR DESCRIPTIONS AND INSTRUCTIONS

Everyone reading this book is an experienced giver and receiver of descriptions and instructions. Whether the subject is how to tie a bow or a knot; how to complete a homework assignment; or how to write a maintenance manual for a nuclear reactor, the keys to communicating effectively are the accuracy and the accessibility of the information.

In 1981, a leading textbook publisher was held financially responsible for the physical injuries sustained by two students while following written instructions for a lab procedure. Inadequate descriptions and instructions can and have resulted, not only in the production of faulty products, but also in the injury and death of persons making or using those products. More often than not, however, descriptions and instructions are accurate, but are misunderstood by the readers. Because the writer has not adapted the material to the capabilities and perspectives of its users, the information is not absorbed as intended.

In fulfilling your daily responsibilities, most of you will write two types of descriptions: a) descriptions of objects, and b) process descriptions that either (1) tell how something works; or (2) tell the readers how to make it work or keep it working themselves.

DESCRIPTION OF OBJECTS

As a technical communicator, the objects you describe most often will be machines and/or parts of machines. Usually, you will be describing these objects as a preliminary step in a process description or so that your readers can use these objects in performing a service. Very often the same objects are described to different people for different purposes. To select an order of presentation and an appropriate language, and to determine how much detail and which details to include, the writer must know who will read the description and how they will use it.

Example 1: A Dentist and His Equipment

A dentist might be required to describe his office equipment on several occasions, each time to different people for different purposes.

Audience 1: Interior Designer

The dentist is renting office space in a new building. The architect has agreed to divide the space as the dentist wishes and to provide consultation with an interior designer to coordinate the dental equipment with the available space. So that the architect and interior designer can do their job, the dentist will need to supply descriptions of his equipment, chairs, X-ray machines, drills, and so forth in terms of their size and shape and in terms of what they do.

The designer audience will not need to know how an X-ray machine works so that they can become dental hygenists. Instead, they will want to know how each machine functions so that they can determine where to put it and what to put near it in the given office space. The details included in the description for the designer are selected based on the writer's understanding of how the descriptions will be used.

Whenever possible, interview the intended readers (or a representative sample) to find out the kinds of questions they need answered. If it is not feasible to speak with them directly before writing, try to imagine the question-and-answer interchange that might take place between you and the reader. An actual or imagined interview before writing helps you anticipate your readers' requirements in designing your description. For example:

DENTIST: I've decided to use mobile units wherever possible. That will give us greater flexibility in treating individual patients.

DESIGNER: You're not going to use a pedestal-type basic unit, then?

DENTIST: No, that one has a handpiece module that moves both vertically and horizontally, but the evacuation syringe and cuspidor systems are stationary.

DESIGNER: Isn't there a single mobile unit that contains a movable tray arm as well?

DENTIST: Yes. But the one I'm ordering suits my purposes even better. It's called a split mobile unit. It divides the instruments that I usually use and the ones that my assistant usually handles between two separate tables on castors.

DESIGNER: I see. In other words, you want to be able to move around the patient and take your equipment with you as you go.

DENTIST: Yes, exactly. But that means I'll need plenty of free floor space around the dental chair.

DESIGNER: The manufacturer's brochure gives the exact measurements of these two units. We'll calculate enough space between the wall cabinets and the chair for free movement of the units. If necessary, we can use narrower, but taller cabinets to give you more floor space.

DENTIST: Sounds good. (Pause) Say, don't forget the analgesic apparatus.

DESIGNER: The what?

DENTIST: The gas machine. It's got lots of precarious hanging things like tubes and cylinders, and patients don't like to look at it. It makes them nervous.

DESIGNER: We can probably put it behind a screen in this corner that faces the back of the dental chair.

DENTIST: Probably so. Now, about the ultraviolet activator light. . . .

Audience 2: The Internal Revenue Service (IRS)

The dentist needs to give the IRS a full description of his equipment to be able to take a depreciation allowance on it. Here is a conversation between the dentist and his accountant that anticipates the different vantage point the writer would have to accommodate in a second description of the same equipment:

TAX ACCOUNT (TA): What about this split mobile unit, Harvey. Have you got the bill of sale? O.K. Let's see, it costs $5000 plus tax. What would you say its longevity is?

DENTIST: Under normal use probably 20 years.

TA: Does that go for all the parts?

DENTIST: I hadn't thought of that. I'd say the work surface, syringe, handpieces and controls, the instrument drawers and storage compartments would all make it that long. But the mixing slab module and the evacuating system—corrosion might be a problem there. Still, I ought to be insured for that. Let's depreciate the whole thing at the same rate.

As you see, the stress here is not on shape, use, or function except as they affect price and longevity.

Audience 3: Dental Students

Because this dentist is using the latest equipment, he might be asked by an instructor at a nearby dental school to give some advanced students a tour of his office. This description would take into account the special concerns of a student audience:

DENTIST: "Experimentation in dental procedures has resulted in the general adoption of seating of the dentist and the assistant during dental operations. That way, more efficient use can be made of the chairside assistant. But to do it right, we needed to replace the fixed-pedestal type dental unit. There are lots of choices: movable pedestal-types, units that attach to the dental chair, units that are built into console cabinets, or mobile units, single or split.

I chose the split mobile unit because of all you can do with it. It's got flexible tubing containing the utility lines, which can be attached to a central outlet with a master switch. That makes the chance of accident or injury during operating minimal. And the assistant and I can move ourselves and our equipment around the patient quickly wihtout having to stand up. In cases of excessive bleeding or procedures where quick adjustments are necessary, we both have it all at hand wherever we are.

Take an ostectomy, for instance. While I'm arranging the soft-tissue flap using the instruments on my unit, the assistant can move around to perform the proper aspiration of saliva and blood without damaging the flap. And we're not both grabbing at the same tray or machinery.

Notice that this time the emphasis is on the medical value of the equipment. Because he is speaking to dentists -in- training, his explanations are much more technical than in the first two instances. Also, since much of his equipment incorporates recent innovations, the dentist makes frequent comparisons to older models.

Audience 4: Fellow Dentists

If the dentist were preparing a presentation for his peers—established dentists at a meeting of the American Dental Association, for example—he would probably put even greater emphasis on comparison with older procedures. When you know that your audience has a particular frame of reference, you can use it to help them understand your innovations. For example, the dentist might say to fellow dentists:

> The new double mobile unit used in combination with the new slim-back-style chairs eliminates all the awkwardness, especially when performing root canal procedures. We used to have to stand up over the patient and lean across each other to get at instruments quickly. Now, you can get really close to the patient and the instruments, while both you and your assistant are comfortably positioned.
>
> You can even remain seated. And, for those of you who think standing up is better for the heart, I refer you to an article in the spring issue of the *New England Medical Journal*. It says dentists have a high rate of heart attacks because they stand too long with their arms in awkward positions that interfere with blood circulation.

Audience 5: Police or Insurance Investigators

If the dentist's office were broken into, he would have two other kinds of descriptions to write of his equipment: a police accident report, and an insurance claim. Again, he would design the description based on the special concerns of the specific audience.

Example 2: A New Product Description

Sometimes, you will write a technical description to familiarize potential users with a new product. Suppose, for example, you were responsible for introducing your projected market to a new rustproofing agent, Zincrometal. Would it be sufficient to provide a list of Zincrometal's properties, such as the one in Figure 10.1? While this list contains almost everything that anyone would want to know about Zincrometal, it puts the burden on the individual reader to pick out the facts that pertain to his or her interests and to apply them accordingly. It is impractical to ask for so much patience and ingenuity on the part of a busy reader. Instead, you can do the bulk of your readers' job for them by following the description structure presented in Chapter 5. Begin your description by providing

1. the reason why the reader needs or would want to understand this description
2. a summary of the main features and why they are relevant to the target readers.

Then, give the pertinent details, clearly. In Figure 10.2 this format is applied to a description of Zincrometal for potential users.

ZINCROMETAL

Used for both unexposed structural parts of vehicles and exposed skin parts

Developed by Diamond Shamrock Corporation

No detectable interface layer between the Dacromet basecoat and the Zincromet topcoat

0.4 mil required in most applications

Contains zinc particles, fine in size and spherical in shape

Projected to increase trade-in value of commercial and private vehicles

Made insoluble by baking at elevated temperatures

Decreases need for rustproofing and maintenance after purchase of vehicle

Basecoat, Dacromet, is a combination of chronic acid and zinc dust

Zincromet undercoating is weldable

Dacromet and Zincromet do not require an intermediary adhesive

Total dry film thickness is approximately 0.5 mil

Van der Wade-type attractive forces are suspected to affect bonding

A hybrid-like system which exhibits synergism

Corrosion-resistant

Dacromet thickness, measured by weight, is 0.1 mil

Marketed through a licensing arrangement with steel producers who maintain their own coating facilities

Independent roll-coating companies may be subcontracted

Figure 10.1   Description list.

PROCESS DESCRIPTIONS

Descriptions of processes or procedures are usually intended for one of two possible audiences:

1. People who want or need to understand how something works.
2. People who must perform the operation themselves.

Often, two versions of the same material will be written—one for each of the two kinds of readers. To understand the different considerations involved in writing descriptions for these two distinct groups, compare the two short essays in Figure 10.3. Both describe the "6-2 offense" in volleyball. The first explanation is aimed at game viewers; the second, at game players. Therefore, in the first version attention is focused on what the onlookers need to know to understand what they are watching, while in the second version the focus is on what someone who will participate in a "6-2 offense" needs to know to implement the recommended strategy.

ZINCROMETAL—THE NEW RUST RESISTOR

Zincrometal is a corrosion-resistant, two-coat system developed by the Diamond Shamrock Corporation to prevent metal rust. It is especially effective on both the unexposed structural parts of motor vehicles and the exterior skin parts. Tests show that when parts are coated with Zincrometal, vehicles have longer life, lower repair costs, increased trade-in value and greater safety. Given reasonable use, automobiles and trucks coated with Zincrometal do not require additional rustproofing after purchase. Marketing surveys show that the average consumer would be willing to absorb a small increase in the initial purchase price of a vehicle in exchange for a reduction in preventive maintenance costs afterwards.

Composition

The basecoat of the new coating process is Dacromet, a combination of chronic acid and zinc dust. Dacromet is 100 percent water based and essentially resin free. It is applied to clean, bare steel and converted into an insoluble coating by baking at elevated temperatures. The undercoat is Zincromet, a weldable, zinc-rich coating. Zincromet is composed of a high loading of fine-particle-size zinc dust in a linear epoxy resin. Both coats are organic solvent based and require baking. Both contain zinc dust particles that are extremely fine in size and spherical in shape. They have a natural attraction for each other and form a strong bond without an intermediary adhesive. There is no detectable interface layer between the basecoat and topcoat.

Coating Specifications

The standard product requires a coating of approximately 0.4 mil of Zincrometal, yielding a total dry film thickness of 0.5 mil. The thickness of the Zincromet may be varied, depending upon the end use requirements, but 0.5 mil has proved to be optimum for most automotive applications.

For Further Information

By calling one of the numbers listed below or filling out and mailing the attached postcard, you can arrange for a Zincrometal representative to visit your company and discuss possible rustproofing applications for your products, at no obligation.

Figure 10.2   New product description.

How a Process Works

In business and technology there are many processes people do not perform themselves which they want or need to understand. The office manager or purchasing agent for a company may not do any word processing but must know how the word processor he is considering for purchase works. The efficiency expert called in to consult with a company on how to reduce man-hours needed

VERSION 1

(This essay is for someone who has played recreational or "backyard" volleyball and has an understanding of the fundamental rules and methods of scoring.)

When a volleyball team uses what is referred to as a "6-2" offense, the action on the court may appear rather chaotic, especially to an observer who is not familiar with volleyball strategies. One player runs from the back of the court to the front of the court, may hit the ball, and then run back to her original position. Meanwhile the other players change position in adjustment to that one player's actions. Given a little background knowledge of offensive strategies, however, anyone can understand the purpose of this seemingly disorganized behavior. A general knowledge of offensive strategies will be helpful.

A team is said to be "on offense" when the ball is on its side of the net. The offensive team wants to efficiently use its maximum of three hits to attack (hit the ball as hard as possible at) the opposite team (the defense). A hard downward hit onto the opponent's court using one hand is called a spike. A person who spikes is known as a spiker or hitter. Only players in the front row may spike the ball. Players try to pass the ball to the hitters so that an attack (spike) can be made. They "set" the ball up for the attack. Any player can set the ball from anywhere on the court, but an experienced team will usually have players that specialize in this activity. They are called "setters". The team wants the setter to be relatively close to the spiker he or she is setting. Normal positioning is shown in Figure 1.

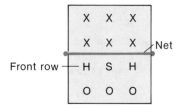

Figure 1   Normal positioning for hitters (H) and setters (S).
X = defensive players, O = other offensive players.

In a 6-2 offense a team has six hitters and two setters. When a setter is in the back row he or she functions as a setter. When that player is in the front row, he or she becomes a hitter and the other hitter/setter comes out of the back row to set. The setter must stay in his/her back row position until the ball is served. After the serve, that player can move to the net, as shown in Figure 2.

As can be seen from the diagram, the advantage of using this offense is that the team will have three hitters available in the front row. The setter has the option of setting any of the three people in the front row, which leaves the defense guessing where the spike will come from. After the offensive play has been made, the setter returns to the back row and her team is now the defense.

To the casual observer, the behavior of the setters running to and from the back court may make the players appear confused when in fact they are using a strategic 6-2 offense, and the seemingly chaotic behavior is well planned.

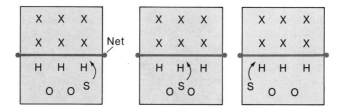

Figure 2   Setter coming out of back row.

VERSION 2

(This essay is directed at someone who has played on organized volleyball teams before, knows the rules and terminology of the game, but is not familiar with playing a 6-2 offense.)

The 6-2 offense in volleyball allows the flexibility and unpredictability of having a middle hitter, but also requires that each player knows the responsibilities of each position on the court.

The setter will always be coming out of the back row. This has several implications: 1) the passes to the setter must be very accurate; 2) the setter has the opportunity to set any one of three hitters; 3) the setter should never receive the serve, and 4) the setter cannot spike or block. The action of the setter coming out of the back row and the relative positioning of the other players is shown in Figure 1.

Figure 1   X = Defense, O = Offense, H = Hitter,
S = Setter

The movement of the setter may lead to potential problems. If the setter cannot get back to her defensive position quickly enough, the two other back-row players must be ready to cover for her. Also, if the setter cannot get into her offensive position fast enough, another player (preferably the other hitter/setter) must be ready to set the ball or at least return it.

The 6-2 offense can be confusing, but most confusion can be avoided by precision passing and correct player positioning. If run properly, it has the potential of providing an aggressive offensive that the opposing team will find difficult to predict in respect to which hitter will be attacking.

Figure 10.3 Volleyball instructions: two versions.

to manufacture its products may never load flash cube trays or operate oxygen furnaces herself, but she will need to know how these processes are performed in order to suggest the best means of reducing labor requirements.

When your job requires you to describe a process to someone who will not actually perform it, precise step-by-step details may not be necessary. Understand how your readers intend to use the information, and provide only what they need to fulfill their intentions.

Figure 10.4 is a description of the Stratified Charge Combustion Concept used in a Honda car. It is intended to explain to a new owner how the car uses fuel. This type of process description aims at assisting the person who will operate the vehicle in understanding why it behaves as it does. Such a description might very well appear near the beginning of a driver's manual and would help justify the operating and maintenance procedures recommended throughout.

Notice that explanations are adapted to a lay audience. For example, in the second paragraph an analogy to a process anyone might be familiar with is used to explain how an engine mixes gasoline with air.

How to Make a Process Work

If you are or expect to be involved in writing descriptions of processes for the people who will perform them, a good way to understand and practice the requisite skills is to write instructions for a game. Choose a sport or boardgame with which you are very familiar and which is complicated enough to require detailed explanation. Assume that your readers are vaguely aware of the process but have no direct experience with it.

First, you will have to think through the steps and strategies to select the facts and to remind yourself of the end goal. Then, you will need to reconstruct the game process to make it clear for your reader.

You will discover that a good process description requires:

1. An opening explanation of the scope and goals.
2. A complete explanation of the objects or materials needed.
3. A step-by-step description of what to do when.
4. A separate discussion of special cases, exceptions, and so on.

You will also discover that accurately drawn and labeled diagrams are central to technical descriptions, especially for the "hands-on" audience. For example, describing the game, "GO" (Figure 10.5), the writer introduces the materials needed and the goals of the game using helpful illustrations. Each of the rules is explained and illustrated, and the main points of the introduction are reiterated in the conclusion. This description is an excellent example of the integration of verbal and visual explanations. It is also a good example of the effective use of repetition at the end of an explanation. An overview of the process is needed at the beginning to orient the reader, but it also functions at the end to pull together the new information that the reader has acquired.

STRATIFIED CHARGE COMBUSTION CONCEPT

Your new Honda is able to use both leaded and unleaded fuel and gets better gasoline mileage through the use of a new Stratified Charge Combustion Concept. (SCCC)

Here is how it works. To use gasoline, the engine must first mix the gasoline with air. This is done in a way very similar to taking some water in your mouth, drawing your lips tight, and exhaling to form a fine spray of water. In an automobile engine, the gasoline is sprayed from a tube into a stream of fast-moving air.

The gasoline-air mixture next flows into each of the the cylinders where it becomes known as the *charge*. In the cylinder, the charge is compressed and ignited.

The amount of gasoline in the charge is very important. If a relatively small amount of gas is used, we say the charge is *lean*. If a relatively large amount is used, we say it is *rich*. In general, a lean charge will give better gas mileage than a rich one because less fuel is used. This is true only up to a certain point, however. If the charge is too lean, the spark plug will be unable to start the gasoline burning. This means some power is lost and fuel is wasted. The object, then, is to mix the gasoline just lean enough so that it will still burn.

In all auto engines up to now, the leanness or richness of the charge is the same throughout the cylinder. However, Honda researchers found that if they made the area near the spark plug very rich and the rest of the cylinder very lean, they could make an engine run on less fuel and produce the same power. This is true because this *stratified* charge is easy to ignite.

Because the new Honda uses so little fuel, it also produces less exhaust. It produces so little exhaust, in fact, that it meets EPA pollution requirements using the "dirtier" (but also cheaper) leaded fuel. The Honda is the only car on the market that can make this claim.

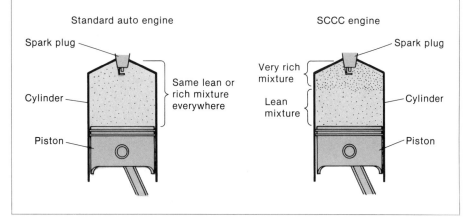

Figure 10.4   Process description.

189

GO

GO is probably the oldest board game in existence. According to legend, it is said to have originated in China more than 4000 years ago. But it was not until its introduction to Japan some 1200 years ago that the game achieved its popularity.

GO demands great skill, strategy, and patience, and is capable of infinite variety; yet the rules and materials are so simple that children can play.

MATERIALS

1. A board with 19 vertical and 19 horizontal lines forming 361 intersections as in Diagram 1:

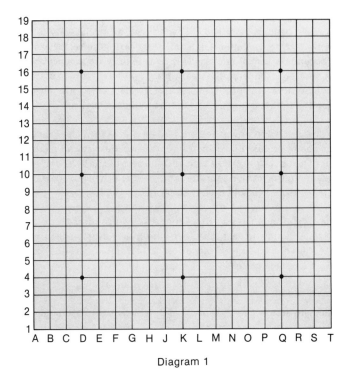

Diagram 1

2. Two sets of "stones" (circular pieces, usually thinner at the edge than the center to allow ease of handling), one dark (Black), and one light (White), enough not to run out (usually about 180 pieces)—see Diagram 2.

Stone

Board

Diagram 2—side view

OBJECT

By judicious placement of his or her pieces, each player tries to control the largest possible territory by making connected enclosing walls, using the least amount of stones.

PLAY

Beginning with Black, each player alternates, placing one of his or her stones on any unoccupied intersection (with one exception—see rule 5). Once placed, the stones are never moved unless captured, at which time, they are immediately removed from the board. The primary purpose of the game is to gain control of territories; the capture of prisoners is quite secondary. Strategic plays have two purposes: to enlarge one's own territory, and to threaten the opponent.

END OF GAME

Play continues until the opponents agree that to continue play is of no advantage to either side. At higher levels of play, a player will resign if he or she sees that his or her opponent has an advantage of more than 10 points.

If there is a difference of opinion as to whether further play may be profitable, the player who wishes to continue may do so while the other passes on his turn. When both players pass, the game is over, and the scores are counted.

RULES

1. Two stones are "connected" if they are adjacent on the same vertical or horizontal line. In Diagram 3, position I, the white stones are connected and the black stones are not.

2. A stone, or a group of stones, is captured and removed from the board immediately when the opponent has occupied every adjacent intersection, to which the surrounded stones connect by straight lines. In Diagram 3, positions II, III, IV, and V, the white stones are captured.

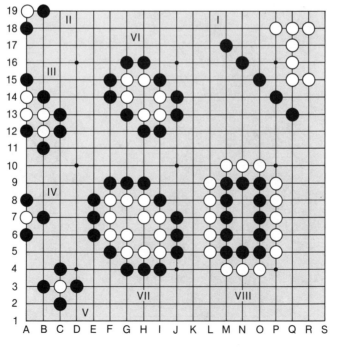

Diagram 3

3. Doomed stones are removed at the end of the game. In Diagram 3, position VI, the white stones are lost. If White places his or her stone on the vacant point H14 the group will be captured on Black's turn and Black can capture the group by placing his or her stone at H14 if White does not. (One might argue that when Black places his or her stone at H14, it is surrounded the moment it is played, and should therefore be considered White's prisoner, but this is not the case; the act of playing H14 and the act of removing the six white stones are both parts of Black's turn to play. Only after he or she has picked up the white stones is it again White's turn to play, and then, of course, White cannot capture the Black stone on H14.)

4. A group of stones with two separate eyes is safe, and counted in the score of the controlling player. An eye is a point surrounded by four stones of the same color. In Diagram 3, position VII, the White group is forever safe; Black cannot completely enclose the White stones in a single turn. In Diagram 3, position VIII, White can eventually capture the black stones by placing a white stone on N7; if it is Black's turn, he or she can place a black stone on N7 and make the group safe.

5. A stone that on the preceding play has captured one stone cannot immediately be retaken if such a capture leads to a repetition of the original situation. Thus, in Diagram 4, position VI, if White plays point E4 and captures the black stone on D4, Black can retaliate by playing a black stone at D4 to capture the white stone on E4; the situation is repeated endlessly without Rule 5. (One might feel that this situation is trivial, but often the life or death of large groups of stones may depend on who finally controls the disputed intersection.)

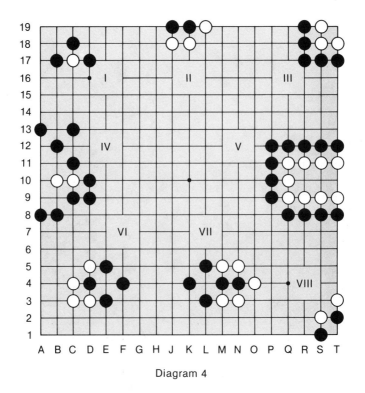

Diagram 4

6. Local stalemate occurs when neither player can dominate (capture) the other. Diagram 5 illustrates a type of local stalemate; if Black plays either J2 or L2, White can capture the group by playing the other point. If White plays either of the points, Black can capture by playing the remaining point. Neither dares to play here. At the end of the game, this area of the board would not be counted as being controlled by either Black or White.

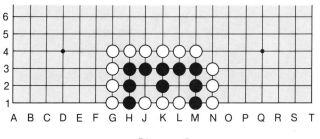

Diagram 5

COUNTING THE SCORE

There are several ways of counting the score, each acceptable:

1. The total score of each player is the sum of all his or her controlled vacant intersections plus any captured stones of the opponent.

2. The total score of each player is the sum of all his or her controlled vacant intersections minus any stones lost to the opponent.

3. White places any captured black stones inside the Black territories, and Black in turn places any captured white stones on any vacant intersections that White controls. Then, each player's score is the sum of his or her controlled vacant intersections.

The winner is the one who controls more intersections (territories).

In Diagram 6, White has enclosed 31 points; namely, A10, B9, M12, N13, and the 27 vacant points on the southern half of the board, between lines C, K, 1, and 6.

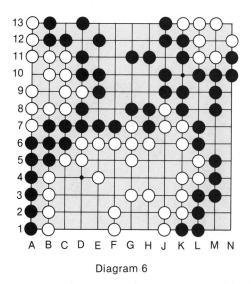

Diagram 6

Observe that the edges of the board serve as natural boundaries of the adjacent areas and need not be occupied to gain control, although each group along the borders must be built down to the edge. Thus, if there was no white stone at B1, the white territory would not be sufficiently enclosed.

CONCLUSION

The rules of GO are few and simple; however, their application can lead to an infinite variety of strategies. The game demands skill, subtlety, and patience. If one is too ambitious, and attempts to build too large a territory, the opponent will be able to play inside without being captured. If one is too cautious, one may be safe, but too small to win. This balance of daring and caution is the crux of the game.

Figure 10.5 Game instructions.

WRITING USER AND MAINTENANCE MANUALS

Learning to explain a game to someone who is less experienced at it than you are is excellent preparation for writing instructions in user and maintenance manuals. Whether the manual explains how to depreciate equipment for tax purposes; how to use a computer, or how to repair a machine part, it is almost always written by someone who, in theory and/or practice, knows the material better than the reader.

As the expert, you must not only include all the necessary information in the manual, you must also format that information for a perspective distinctively different from your own.

In a well-written user manual:

1. The introduction includes a clear description of audience level: "it is assumed that you are a novice, but you understand. . . ."
2. The table of contents reflects the user's, not the writer's interest in the material.
3. All terms are defined and all materials are described *before* they are discussed/used.
4. Examples of each incident or function are given.
5. Paragraphing is at a minimum and listing, indentions, and subject headings are used to make comprehension easy.

(For a good example, see the "User's Guide to the Dec 20" in Appendix C.)

The excerpt from Motorola's Instruction Manual for the Intrac 2000 Central Station (Figure 10.6) exhibits similar characteristics, even though it is addressed to a more expert audience ("This manual is intended for use by experienced technicians familiar with similar types of equipment," p. IV). As illustrated by

DECODER MODULE
MODEL CLN1120A

1. GENERAL

The CLN1120A Decoder Module is a plug-in circuit board attached to the Alarm Panel. The board contains the audible alarm buzzer as well as the logic circuitry, and is electrically connected to the OVERFLOW indicator and two ALARM RESET pushbuttons on the panel.

1.1 The primary functions of the decoder are decoding signals received via the communication channel, and updating the information displayed on the rolling and fixed display modules.

1.2 The audible alarm is activated either by a COS detection in a display module or by an overflow of the rolling display.

2. GLOSSARY

The mnemonics, signals and terms defined herein appear in the diagrams and discussions included in this section. Please note: a bar ($^-$) over a signal name denotes the complementary function of that signal.

$\overline{ACK}$ *(ACKNOWLEDGE):* A signal activated once the ALARM RESET: VISUAL pushbutton on the alarm panel has been depressed; it halts the flashing of the address indicators on rolling display modules, where applicable.

BCL (BASIC CLOCK): An output of the clock generator, this signal clocks the control circuit.

BZO (BUZZER ON): A signal from the display control, it is logic 1 whenever the audible alarm sounds; once it has been reset by the ALARM RESET: AUDIO pushbutton on the alarm panel, it goes to logic 0, indicating that the alarm is quiescent.

COM (COMMON): The current return path for power devices; e.g., drive transistors, buzzers, etc.

$\overline{COS}$ *(CHANGE-OF-STATE):* An input from one of the display modules indicating that at least one of the status/alarm bits (16-23) of the code-word does not match the information previously stored in the module.

DAI (DISPLAY AVAILABLE INPUT): An input from the shared display indicating the presence of at least one unoccupied rolling display module.

DATA: The received data output of the data sensing circuit, it contains all 32 bits of the received code-word.

DOT (DATA OUTPUT): The valid data output of the decoder, it is forced to logic 0 during rolling, thereby inhibiting data transfer from the decoder to the shared display until after rolling has been completed.

DUP (DISPLAY UPDATING): A signal indicating that period during which data is transferred either from one rolling display module to another by rolling or from the decoder to a display module once the decoder has determined the validity of a received word.

(M) **MOTOROLA INC.**

service publications

Figure 10.6 Excerpt from a user's manual. (Courtesy of Motorola Inc., Communications Division, Schaumburg, Illinois.)

ECNT (END OF COUNT): A signal from the timing circuit to the control circuit indicating either that the timing circuit has reached count 39 or that REJECT had been generated during data reception.

$\overline{FCL}$ *(FAST CLOCK):* A clock generator signal applied to the display module during updating, it clocks the data bits of the received code-word into the registers of the intended display module.

$\overline{FLR}$ *(FLASHING RATE):* An oscillated signal generated by the display control, it activates flashing of appropriate indicators on the display modules.

FSR IN (FSR INPUT): The FSK tone input of the received signal, applied to the FSK receiver from the line interface unit.

FSRO (FSR OUTPUT): The FSK output from the FSK receiver to the clock generator.

FSS (FRAME SYNCHRONIZATION SENSING): A data sensing circuit output which goes to logic 1 at the end of a received word, thus differentiating between consecutive words.

GND (GROUND): The current return path for the logic circuitry only.

ICL (INPUT CLOCK): A pulse generated upon each transition of the pulse duration modulated output (FSRO) of the FSK receiver: the time interval (T.2T) between consecutive pulses determines the logic of the respective bit (0.1).

NAD (NEW ADDRESS): A signal that remains at logic 1 at the termination of PH1 if the address portion of the code-word does not match an address currently stored in any of the display modules of the system.

NCH (NEW CHANGE): A signal from a display module, which, if logic 1, activates the audible alarm to indicate the detection of a change-of-state; if forced to logic 1 by the decoder, it indicates a display overflow.

OVF (OVERFLOW): A signal from the control circuit indicating that no display is available for a received message; it causes activation of the audible and visual alarms on the alarm panel.

OVR (OVERFLOW ROLL): A signal from the control circuit, it rolls the entire shared display one module to the left once an overflow condition has been detected. (Not operational.)

PARITY: The parity check signal from the validity detector indicating, if logic 1, that an odd number of logic 1 bits is contained in the received code-word, a condition necessary for a valid code-word.

PH1 (PHASE 1): A signal generated by the control circuit whenever the 13 received address bits (0-12) are shifted out of the basic register in order to be compared with addresses stored in the display modules; it lasts until the address comparison has been completed.

$\overline{READOUT}$*:* A signal (logic 0) which is generated by the control circuit only in instances where DUP is concerned with data transfer to display modules directly from the decoder (as opposed to rolling).

REJECT: A signal from the data sensing circuit indicating that one or more received bits is less than 0.5T duration.

RESET: A pulse that resets all registers and counters either once FSS has been detected by the control circuit or a START pulse has been generated upon power-up.

ROLL: A signal to the rolling display modules; if logic 1, it initiates the rolling cycle and sets the timing circuit to count 13.

RRQ (ROLLING REQUEST:) A signal which, if logic 0, indicates that a condition requiring rolling exists in the shared display.

SHIFT: A signal that shifts the data in the basic register either during reception of a message or during DUP.

START: A pulse from the power supply; upon initial power application or restoration, it causes all registers to be reset to their initial states.

STR (STROBE RATE): A display control signal required in the rolling display modules for the multiplexing of the station address indicators.

TCL (TIMING CLOCK): A pulse applied to the timing circuit: it is generated by either ICL during data reception or $\overline{FCL}$ during DUP.

TRN (TRANSMIT): An encoder-generated signal that disables the decoder while the encoder is transmitting.

VALID: A signal generated by the validity detector to indicate that the code-word has met all of that circuit's security checks.

Figure 10.6 *(Continued)*

V_{IN} *(INPUT VOLTAGE):* The input voltage (14.5 V dc) provided by the self-contained power supply (TPN1136A or TPN1154A).

V_{LG} *(LOGIC VOLTAGE):* An internally generated voltage (5.6 V dc) for the logic circuitry.

3. FUNCTIONAL DESCRIPTION

The decoder has two primary functions: the decoding of signals received from remote stations; directing data updating of the display modules.

3.1 DECODING

The decoder accepts FSK (frequency shift keyed) signals from the communication channel (via the line interface unit) and converts them first to pulse duration modulated signals and finally to a binary code-word which it checks for validity before passing it to the display modules and/or printer.

3.1.1   The FSK receiver accepts the FSK tones (900 and 1500 Hz) from the line interface unit and converts them to pulse duration modulated (T, 2T, 4T where T = 1.67 ms) signals before passing them to the clock generator for processing.

3.1.2   In addition to producing all of the clock signals necessary to operate the decoder, the clock generator detects transitions between the pulse duration modulated signals, generating SHIFT pulses at each transition which are then passed to the data sensing circuit for decoding.

3.1.3   Data sensing determines the time interval between consecutive SHIFT pulses, converting each of them to a binary code according to the following criteria: 0.5T to 1.5T = logic 0; 1.5T to 3.0T = logic 1; >3.0T = frame synchronization (FSS). Any bit received as less than 0.5T causes rejection of the entire code-word. The SHIFT pulses now pass the decoded data bits in parallel to both the basic register and the validity detector. Note that only 26 SHIFT pulses are generated for the data intended for the 26-bit basic register inasmuch as the security bits have no further application in the data updating which follows.

3.1.4   The timing circuit counts the sequential number of each bit shifted in or out of the basic register at any given time. Should a cause for word rejection (underflow, overflow) be detected, the circuit generates an end-of-count

signal to the control circuit, thereby terminating the updating cycle.

3.1.5   The validity detector, together with the BCH and PCH circuits, performs three security checks necessary to validate a received word: Bose-Chaudhuri (BCH) cyclic code which compares received bits 26-30 with a decoder-generated code wherein each code bit is based on all of the previous bits; parity check (PCH) which checks for an odd number of logic 1 bits; and bit count (32 bits longer than 0.5T duration). Only after all of these checks have been determined to be correct is a VALID signal generated and passed to the control circuit, thereby enabling the updating cycle to follow.

3.2 DISPLAY UPDATING

After the received code-word has been validated, display updating takes place. This includes updating of individual display modules as well as the collective updating of the shared display, i.e., the alignment (by rolling) of all occupied modules contiguous to the left.

3.2.1   The control circuit determines the decoder's mode of operation as well as the status of each display module, according to the validity of the received word. If the word is valid, display updating begins with the decoder making an address comparison in order to determine which module will be updated by the received message. If the word is not valid, updating is disabled and the decoder is ready to receive the next message.

3.2.2   During updating, the contents of the basic register are shifted out and looped back into the register's input. During a rolling operation, the register circulates the contents without accepting new information; if, however, the code-word's validity had been established prior to rolling, an additional updating cycle enters the valid data to a display module once rolling has been terminated.

3.2.3   Rolling occurs once a rolling display module advises the control circuit that there is at least one vacant display to its left, but not before the decoder has finished processing the current received message.

3.2.4   The display control circuit interfaces the decoder with the alarm panel and display panels. It activates and resets the audible alarm and flashing of the status/alarm and/or address indicators on the display panels.

Figure 10.6   *(Continued)*

4. DETAILED CIRCUIT DESCRIPTION

Refer to the decoder module schematic diagram.

4.1 FSK RECEIVER

This circuit accepts FSK signals from remote sites via an rf or wire line communication channel, and converts them to digital words; FSRO is recognized by the decoder as a PDM message.

4.1.1 The FSR input tone at P1-A is converted to a square-wave signal of the same frequency at U8A-3 (TP2) by passing it through the input filter and the limiter.

4.1.2 The square-wave signal is applied to the phase lock loop at U25C-10 which generates a signal of the same frequency, but with a phase shift dependent upon the frequency.

4.1.3 The phase shift is converted to dc level at U10D by the phase comparator and the data filter, consisting of U22A and B, respectively, and all of their associated resistors and capacitors. U22C and D comprise the slicer

while a transition detector, consisting of U10A, C26 and R52, produces an ICL pulse in each pulse duration transition.

4.2 CLOCK GENERATOR

This circuit generates the clock signals required in the operation of the decoder.

4.2.1 ICL (INPUT CLOCK)

This signal is generated from the digital message appearing as FSRO, and is applied to data sensing which processes the data in the message. Each transition (either positive or negative) of FSRO is converted into a positive ICL pulse at U10A-3 (TP11). Thus, the time duration between two consecutive pulses is equal to the duration of the received bit between two transitions.

4.2.2 BCL (BASIC CLOCK)

This signal is the output at U24B-4 of a free-running square-wave oscillator (U24A-U24D, R54, R56 and C27) which is used to clock the control circuit.

Figure 10.6 (*Continued*)

the section discussing The Decoder Module, each part of the equipment is discussed using the following format:

1. *General*
Summary of appearance and function

2. *Glossary*
Definition of terminology

3. *Functional Description*
Summary followed by numbered details under each sub-function

4. *Detailed Circuit Description*

These descriptions are accompanied by two graphic aids: a circuit board detail and a schematic diagram with parts list.

Engineers, programmers, or technical writers who prepare instructional manuals are also frequently responsible for issuing service bulletins or updates to manuals. Figure 10.7 illustrates the use of the Problem/Solution format to add new information to a looseleaf-style maintenance book. Next to "Solution"

Subject: 877 Strip Chart Recorder Pen Redesign.

Problem: Several failures of the bottom pen due to separation of the cement holding the stylus from the metal base plate.

Solution: Both lower and upper pens were redesigned for greater strength. The newer pens can be identified as follows: The older lower pen bracket supported the stylus with a simple right-angle bracket allowing cement to adhere to only 2 surfaces. The newer bracket extends the support up two sides forming a box and allowing the cement to adhere to 4 surfaces. A stylus support was added to the upper pen in the form of a metal strip or band, bent in a U shape around the stylus. Cement is packed around the stylus, filling the space between the support strip and the bracket.

Action: Service branches should send in any stock they have of un-used old style pens for exchange for new ones. Also, customer's old style pens should be replaced whenever possible, such as when performing other service

It would also be appreciated if you would return any new style 877 pens which fail in the future, for Q.A. evaluation.

Reference: XXXX

Figure 10.7 Excerpt from a service bulletin.

there is a discussion of the difference between the old bottom pen design and the new, but only the old design is illustrated. Could the comparisons have been illustrated more effectively?

The best way to learn what constitutes good descriptive and instructive writing in a lab report, a sales brochure, a textbook, a proposal, or a user's manual is to read critically as many samples of descriptive and instructional writing in your particular field as you can find. In doing classwork or performing your job, be aware of anything you read that works or does not work, and why. By understanding what *you* require as a reader, you can meet the requirements of others for whom you will write.

EXERCISES

1. Write a description of a game (sport, boardgame, etc.) that is complex enough to require some technical expertise. Your reader is someone who knows of the game but has never actually played it. Test your description on members of your class to see if you have met your readers' requirements.

2. The following two examples were written in an undergraduate technical writing class. The assignment was to write a description of the same object or procedure for two different audiences. By yourself or in a workshop group, evaluate each of the two essays for effectiveness of audience adaptation and completeness of description. Do the readers get what they need in a form they can comprehend?

ESSAY 1: BILLING OPERATIONS

Version 1: To an Entry-Level Operations Clerk

This is a description of your daily job routine. It is important for you to follow this guideline so the tasks you are responsible for can be accomplished on time. An overview of the functions you are to perform is listed below:

1. Process local bills
   a. Tear down bills.
   b. Match bills with order tickets.
   c. Distribute pink copies.
2. Have commodity bills sent over B teletype.
3. Prepare teletypes A and B for daily traffic.

The first item you take care of is the local billing. When you arrive at 7:00 a.m. the bills will be in the basket on teletype B. These bills are generated in New York when an order placed by our office is transacted on the stock exchange. The daily transactions are compiled by computer and transmitted to our office in account number order after the market closes. The billing is to be removed from the printer and set aside while you prepare the B printer to receive the commodity bills.

Change from the code 7000 paper to the 3184 commodity billing paper. It is imperative to have the 3184's sent to our office by 8:00 a.m. so bookkeeping can check them for errors before the commodity exchange opens. After the 3184 paper is loaded into the B printer, request a test via the CRT to make sure the paper is lined up correctly. After the paper is lined up, wire New York on the CRT to let them know they may send the commodity bills over the teletype. The billing will be transmitted to the teletype shortly after you wire your request.

While the commodity bills are being sent, you will resume working on the local billing, which was removed from the teletype earlier. The bills are connected to one another on one continuous form and must be separated by tearing along the perforations between them. This task should take 20 to 30 minutes, depending on the number of orders placed the previous day.

The next step in the local billing process is matching the bills with their respective order tickets. The order tickets, which have been filed in account number order, will be on your desk. Compare each bill against its order by checking the following items:

1. Account number.
2. Name of client.
3. Broker number.
4. Number of shares transacted.
5. Name of stcok transacted.
6. Price transacted.

If all items are in agreement, break the bill down to separate the four copies. The brown and white copies get stapled to the order ticket and are placed in a pile for bookkeeping. The blue copy is put in a pile to go to the mail room, and the pink copy is set aside to be given to the broker. Any items that do not agree are to be left intact and placed in the corrections box along with the corresponding order ticket.

Continue matching the orders with the bills until 9:00 a.m. At this time you must change the paper in both teletypes A and B to the code 1000 paper to receive daily messages. Remember to change the tension setting on the teletype printing heads to 1.5 for the thinner code 1000 paper; otherwise, the printers will not function.

After the teletypes are set up to receive daily traffic, finish with the local bills. You should complete matching the trades no later than 9:45 a.m. The matched brown and whites are to be taken to bookkeeping, the blues to the mail room. The pink copies are to be sorted according to broker number and then distributed to the brokers by 10:10 a.m. (before the exchange opens) so they may check for errors you might have missed or that are a result of an incorrectly written order.

In summary, you should have the commodity bills sent by 8:00, the teletypes prepared to receive daily traffic by 9:00, and the local bills processed (pink copies distributed) by 10:00.

Version 2: For an Operations Manager

This is a description of the duties to be performed by the local billing clerk. An overview of the clerk's functions is listed below:

1. Process local bills:
    a. Teardown.
    b. Match orders to bills.
    c. Distribute pink copies.
2. Have commodity billing sent over B teletype.
3. Prepare teletypes A and B for daily traffic.

The clerk's first task should be to remove the local bills from the B printer and change to commodity billing paper.

The second job is to request the 3184's from New York by 8:00 a.m. After the 3184's have been wired for, the clerk should continue processing the local bills.

The third item the clerk is responsible for is loading the A and B printers with code 1000 paper to receive daily traffic by 9:00 a.m. When the paper has been changed, the clerk should finish the billing and distribute the pink copies to the brokers by 10:00 a.m.

ESSAY 2: DESIGN OF ALUMINUM DIE CASTINGS

Version 1: For High School Students

Aluminum die (high-pressure) castings compete directly against molded plastic or fabricated (assembled from stock items) components in many applications. The casting process, injection of molten aluminum into a hardened steel mold, serves high-volume requirements almost exclusively. The high initial cost of the mold is best offset by distribution over the large quantity of pieces a die cast die can produce.

The high-pressure injection (up to 15,000 psi) allows the forming of a variety of items. Both intricate carburetors and transmission housings can be cast, to illustrate the spectrum of application.

The die-casting process eliminates many machining operations because of the consistent high tolerances provided by the permanent tooling. In contrast, other methods of casting aluminum use "perishable" tooling. Perishable tooling includes sand molds formed by a master pattern for only one use. The forming of the sand and distortion in contact with the molten aluminum require additional wall thickness and excessive machining to create identical parts. The dense casting structure created by high pressure injection yields strength and light weight.

The mold splits at a parting line to allow casting removal. All surfaces perpendicular to the parting line must be tapered to allow the piece to release from the mold. An insufficient degree of taper will cause the casting to seize in the mold with possible damage to the cavity.

The cavity, or configuration responsible for the shape of the casting, is a mirror image of the piece it produces. The cavity is created in reverse of the part drawing through normal machine tool operations, basically by milling and hand polishing to achieve a smooth finish.

This primer to casting design would best precede a visit to a foundry or mold-making shop. The myriad applications for die castings and their specific designs would be further explained by visual examples of a part and the tool that produces it. The close inspection of an automobile or home appliance would also lend much information about design and utilization of castings.

Version 2: For a Product Designer

The ultimate configuration of a cast aluminum part is a compromise between cost-effectiveness and design flexibility. Subtle differences can affect piece and tool price as well as casting performance.

The intent of this outline is to inform the designer of basic considerations he must take in developing the initial part concept.

A. Wall section can be maintained at a .06 inch minimum for most cosmetic applications. High-stress areas, such as screw for bearing bosses, require a 3:1 wall stock increase over the chosen minimum section. All changes in wall section should be accomplished through generous radii to eliminate stress concentration. Drastic wall section changes should be avoided, utilizing radial or tangential ribs to support nonsymmetrical or protruding features. The foregoing guidelines are useful in determining minimum material cost.

B. The configuration of the parting line (PL), or seam between cavity halves, must be determined, as draft is required on all surfaces perpendicular to the PL. Refer to the American Die Casting Institute (A.D.C.I.) manual for specific draft tolerances.

C. Alloy selection is determined through intended usage. Variables to consider include: (1) required tensile and yield strengths, (2) need for corrosion resistance, (3) final finish, (plating, polishing, etc.), and (4) degree of secondary machining operations. In turn, alloy selection also affects casting integrity and production rate. See the A.D.C.I. manual for alloy specifics and properties.

The foregoing items do not constitute an all-inclusive summary of design considerations. They are intended to place the initial design in a middle ground, allowing further refinement through contractor's input at the preliminary quotation stage.

CHAPTER 11

Writing Proposals

The well-written proposal:

- Begins with a clear and thorough opening summary.
- Addresses the major concerns from the intended readers' point of view.
- Anticipates and answers possible questions and objections.
- Uses headings, paragraph structure, and language that increase ease of understanding.

To plan a well-written proposal:

- Assemble your facts.
- Read over your facts to establish a priority message and divide your supporting information into categories.
- Assemble a list of readers' actual or supposed criteria.
- Analyze the assembled information on your subject and your readers to determine an effective strategy and format.

ANTICIPATING YOUR READERS' CRITERIA

When you are writing a bid for a contract or a proposal for a new product, or a change in procedures, you are in a situation where the pursuit of your work depends on the consent and cooperation of others. More often than not, the end goal of these others for whom you are writing a contract bid or proposal is the same as yours. If you are in the same organization, you each want that organization's products or services to be well-received by clients and customers. If you are in different organizations, both you and your readers are motivated to cooperate with groups and individuals who can improve the products and services you handle and/or assist the business or consumer community as a whole.

But, no matter how many common goals writers share with their readers, they may not share the same perspective on the merits of this particular proposal. Their criteria for deciding on a specific plan may be very different from yours, even when you are all working towards the same end.

Unless you know what criteria your readers will apply to your suggestions, you cannot present your proposal in a format that anticipates readers' questions and, as much as possible, incorporates the answers into the proposal. A well-written proposal leaves *no* question unanswered that would make your readers hesitate to accept your plan after they have read it.

The best way to be sure of readers' criteria is to get them in writing directly from the source. If you are bidding for a government contract, or requesting the financial support of a foundation, chances are that the readers will provide detailed written instructions. If not, you may be able to interview the key readers in person or by phone. If they have not provided written criteria, you write up a summary of the guidelines mentioned in the interview, and mail them a copy, asking them to confirm your undersanding of their requirements.

If you have no written guidelines, and time or convenience prevents personal contact with the principal readers before submission of the proposal, the next best procedure is to suppose what their job responsibilities involve and to imagine yourself as someone in their place reading your proposal.

For example, the property of an industrial manufacturing business includes a large, grassy plot at the southeast corner of the complex. It has been unused for several years and is adjacent to two other commercial buildings with several hundred employees. The company offers a prize for proposals on how to use this property. One of the entries suggests that an on-site jogging and exercise course be built for use by employees before and after work and during lunch hour and breaks. The proposer, a plant machinist who happens to be an avid jogger, introduces some important issues from the managerial point of view, such as the fact that (1) there are no parks, shopping areas, or interesting restaurants within a half-hour of the plant, and (2) workers either are late getting back from lunch because they travel too far to find midday meals or recreation, or else, are bored and frustrated by being confined to the plant cafeteria. Furthermore, productivity tends to slow down in the afternoon hours. An on-site exercise facility would cost little, improve morale, and encourage promptness and productivity.

So far, so good. As he reads the proposal, however, the decision-making manager begins to wonder how the construction of the outdoor facility would affect the neighboring commercial buildings. Would the ongoing activity distract workers in their own and in adjacent companies' offices? Would they have to hire a security guard and issue passes to prevent use by workers from other plants? Would they need additional insurance? Would their present resident nurse be sufficient to treat athletic injuries in addition to on-the-job accidents?

While the manager is reading the proposal, these and other unanswered questions about the effect of the suggested exercise facility on the company's liability for its own and its neighbors' personnel weigh heavily on his mind. If he continues to ponder these questions while reading the rest of the proposal, he is distracted and may miss one or more of the machinist's strongest points. As a result, a proposal that might otherwise be acceptable may be rejected.

Most proposal readers have pen and paper beside them to jot down questions or objections as they read. Your goal is decision-making readers who arrive at the end of your proposal with a blank piece of paper and an unused pen beside them. The only thing they should be left wondering is why they cannot think of any good reason not to do what you are suggesting.

PLANNING A PROPOSAL

As with any technical writing task, planning a proposal means accumulating sufficient information about a subject and pertinent facts about the intended readers, and then analyzing both groups of data to structure an effective communication.

Step 1: Assembling Evidence

As in the planning of a memo, your first step is to assemble all the supporting evidence you can find. Lay it out in front of you, in list form, if it is short enough, or in an outline derived from more extensive notes.

Step 2: Identifying the Main Message

Next, read through all your information or the outline summary, looking for the answers to two questions:

1. What is my priority message? What do I want the readers to do as a result of having read my proposal?
2. Which items of information that I am presenting fall under each of the following main categories of a proposal—(a) Problems, (b) Objectives, (c) Methods, (d) Evaluation?

As Norman J. Kiritz suggests[1], dividing your information into these four categories in answer to question 2 may help you arrive at the answer to question 1—the priority message.

Step 3: Identifying Readers' Criteria

Using written instructions, or your own assessment of your decision-making readers' criteria, decide what your main readers' priorities will be in judging your proposal.

[1]Norton J. Kiritz, *Grantsmanship Center News*, no. 6 (Copyright © 1974), 1978, pp. 11–14.

Step 4: Analyzing the Information

Place in front of you

1. Your priority message.
2. Your lists or outline of problems, objectives, methods, and evaluation.
3. Your analysis of your readers' criteria.

Read over this information to determine a method of presentation that will produce the desired result for the given reader.

Step 4 is the hardest part of proposal writing. It requires patience and courage to sort out and process your thoughts. You must be prepared to exercise your mind to its limit; but you must also get away from the material periodically and give your mind time to refresh itself between brainstorming sessions. Your reward is that eventually the right shape for the material will present itself. Once you see the plan, the writing itself will be relatively simple. The sooner you try this method, the better, since seeing it work reinforces your ability to apply it. Keep re-thinking the material and the needs of the audience until an effective means of structuring your proposal presents itself. After your have used this approach a few times, you will find that you develop two or three effective formats, and that one of these will always be adaptable to almost any proposal writing. In other words, the first few times that you spend what seem like endless hours in the planning of a proposal, you are not merely solving your immediate professional needs—you are also learning a process that will make your future proposal-writing tasks considerably easier.

Example 1

Suppose your are connected with a fairly new company that converts biological wastes into methane gas which it sells to utility companies. Specifically, your firm has developed an effective method of transforming cow manure into energy. Your assignment is to write a proposal for Community Gas, Inc. suggesting that your company supply them with 1.6 million cubic feet of methane gas per day at $1.94 per 1,000 cubic feet. You want a contract with them that will be renewable at the end of one year subject to the satisfaction of both parties.

Step 1

In the methane gas example, you would assemble all the facts about how your company makes methane gas out of cow manure. Particularly because this is a new process, you will need to explain it in detail to establish credibility. But while you need to be thorough, you also need to be clear. Your explanation of how your company does the conversion should probably be accompanied by a flow chart. Figure 11.1, for instance, is an easy-to-follow illustation.

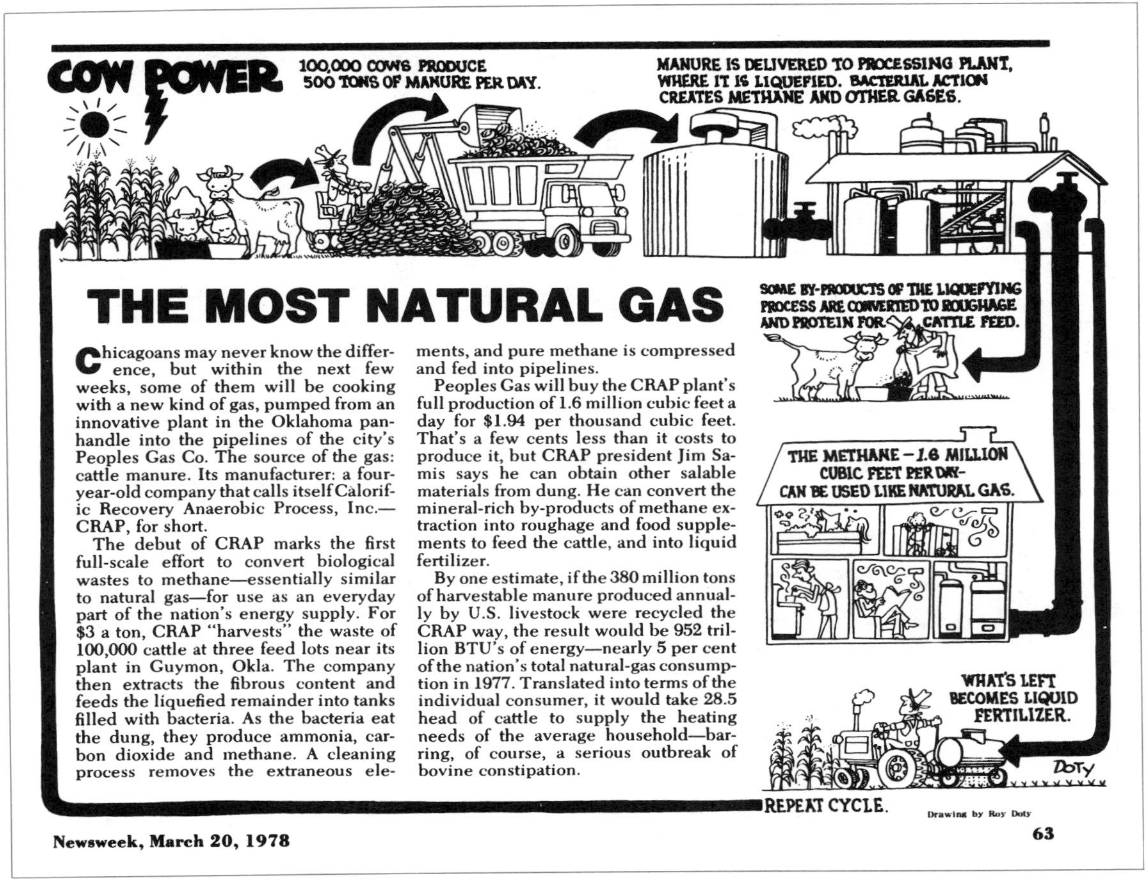

COW POWER
100,000 COWS PRODUCE 500 TONS OF MANURE PER DAY.
MANURE IS DELIVERED TO PROCESSING PLANT, WHERE IT IS LIQUEFIED. BACTERIAL ACTION CREATES METHANE AND OTHER GASES.
SOME BY-PRODUCTS OF THE LIQUEFYING PROCESS ARE CONVERTED TO ROUGHAGE AND PROTEIN FOR CATTLE FEED.
THE METHANE – 1.6 MILLION CUBIC FEET PER DAY– CAN BE USED LIKE NATURAL GAS.
WHAT'S LEFT BECOMES LIQUID FERTILIZER.
REPEAT CYCLE.
DOTY
Drawing by Roy Doty

THE MOST NATURAL GAS

Chicagoans may never know the difference, but within the next few weeks, some of them will be cooking with a new kind of gas, pumped from an innovative plant in the Oklahoma panhandle into the pipelines of the city's Peoples Gas Co. The source of the gas: cattle manure. Its manufacturer: a four-year-old company that calls itself Calorific Recovery Anaerobic Process, Inc.—CRAP, for short.

The debut of CRAP marks the first full-scale effort to convert biological wastes to methane—essentially similar to natural gas—for use as an everyday part of the nation's energy supply. For $3 a ton, CRAP "harvests" the waste of 100,000 cattle at three feed lots near its plant in Guymon, Okla. The company then extracts the fibrous content and feeds the liquefied remainder into tanks filled with bacteria. As the bacteria eat the dung, they produce ammonia, carbon dioxide and methane. A cleaning process removes the extraneous elements, and pure methane is compressed and fed into pipelines.

Peoples Gas will buy the CRAP plant's full production of 1.6 million cubic feet a day for $1.94 per thousand cubic feet. That's a few cents less than it costs to produce it, but CRAP president Jim Samis says he can obtain other salable materials from dung. He can convert the mineral-rich by-products of methane extraction into roughage and food supplements to feed the cattle, and into liquid fertilizer.

By one estimate, if the 380 million tons of harvestable manure produced annually by U.S. livestock were recycled the CRAP way, the result would be 952 trillion BTU's of energy—nearly 5 per cent of the nation's total natural-gas consumption in 1977. Translated into terms of the individual consumer, it would take 28.5 head of cattle to supply the heating needs of the average household—barring, of course, a serious outbreak of bovine constipation.

Newsweek, March 20, 1978

63

Figure 11.1 Flow chart of gas conversion process. (From *Newsweek,* March 20, 1978, p. 63.)

Step 2

You also need to assemble facts to substantiate your claim that there are advantages to using your methane gas as opposed to other sources of natural gas. Can you support claims that it is cheaper, easier to use, more reliable, and more adaptable to different users?

Step 3

Next, you will probably call the appropriate representatives at the potential client firm and arrange to speak with them to ascertain their needs, expectations and questions. Let us assume, however, that for one reason or another information about your readers' anticipated questions or objections is not readily available to you in writing or through direct contact.

Put yourself in the place of your proposal target audience, and see if you can suppose what their concerns might be. The purchasing agent for Community Gas, Inc., for instance, might find some of the following questions entering his mind as he reads your proposal:

- How can they guarantee the delivery of the promised 1.6 million cubic feet? What if the cows get sick?
- Do they have a back-up herd or an emergency supply source?
- Does this gas meet the specifications of the EPA and other government regulatory agencies on whom the licensing of our utility company depends?
- Is there a track record? Are any companies comparable to ours currently using your services?
- Your flow chart tells me how you process the gas, but will there by any change necessary in processing and distribution at our end?
- Will distribution of this type of methane gas affect our insurance coverage? Our union contracts?

Step 4

Now, you will decide on a unified approach for your proposal that (a) presents the information accurately, and (b) addresses the readers' perspectives. For example: "A partial solution to the Community Gas Company's problem of making supplies meet demand without raising its costs and, consequently, its prices, is to supplement natural gas with methane gas converted from biological wastes."

Example 2

You are an electrical engineer in a major manufacturing firm that sells appliances, radios, televisions, and other types of electrical equipment. Some years ago your firm attempted to enter the computer hardware manufacturing market and was unsuccessful. On the basis of observations you have made about changes in marketing situations, computer technology, and your company's product line, you feel that this would be an opportune time for them to re-enter the computer hardware market.

Your preliminary proposal is aimed at your division head, the Director of Research and Development. If the director thinks your suggestion has merit, it will be forwarded to the Head of Marketing.

Step 1

You have read everything available on what happened the first time your company attempted to enter the computer market and you have talked with some of the other personnel who were around then. You are able to pull from your research the following major facts:

* Twelve years ago the first surge in the computer market peaked.
* There was one large industry leader; the profit margin in the industry as a whole was low; and the competitive climate was such that only companies strategically well-positioned in research, pricing, distribution, and services could survive.
* Your company, as well as other large concerns that had put a tentative foot in the door, decided to pull out rather than invest excessive capital in developing a market.

Given the supporting evidence, you feel that, at the time, your company made the right move. However, your proposal is going to argue that they should re-enter the computer hardware market now, because the prohibitive factors that existed 12 years ago are no longer a deterrent. As part of step 1, you need to list the decisive changes that have occurred to create a favorable competitive climate:

* With advancements in the manufacturing of semiconductor and Large Scale Integration (LSI) chips, the design of computer hardware is becoming increasingly standardized.
* Because research and development for computer hardware components is presently undertaken primarily by the chips manufacturer, entering the computer market requires little initial capital.
* Due to projected increases in home and office use of computer hardware, the present and future market is now large enough to be shared comfortably by several companies.

Although your main argument is the documented favorable change in the competitive climate, you also see other advantages for your company in manufacturing computer hardware:

* Your company already carries products such as televisions and appliances that use the same type of distribution channels and services as small computers. Therefore, computer products would fit into the existing product line and strengthen it.

- Computer technology is also spilling over into many products your company already manufactures such as mobile radios, dishwashers, and medical test equipment.
- A major competitor with a similar product mix has already acknowledged its favorable market position by acquiring a major manufacturer of small computers and forming XYZ Data Systems Group.

Step 2

In this case, in the process of listing your information in step 1, you have already accomplished most of step 2. Your priority message is that your company should re-enter the computer hardware market because conditions for doing so are presently very favorable. Your supporting evidence divides into (1) a comparison of market conditions 12 years ago when the company first attempted entry and withdrew, and conditions now when, as you propose, they should try again; and (2) other advantages besides a favorable competitive climate for adding computer hardware to the firm's product mix. This material covers the problems and objectives of your proposal. Since yours is a preliminary proposal suggesting that your idea be accepted, you may not even want to go into methods of entering the computer market and means of evaluating success. These categories may make up a second-stage proposal, once your initial idea is being entertained seriously.

Step 3

Since the suggestion that your company re-enter the computer hardware market is an unsolicited proposal, you will need to make educated guesses about your readers' criteria. Probably, they will be asking the two major questions to which the information you listed in step 1 and evaluated in step 2 is addressed:

1. Does the competitive market that was lacking twelve years ago exist now?
2. Are there any other advantages besides a competitive climate to warrant adding computer hardware to the firm's product mix?

But, they will also be asking a third question that the material you have collected so far does not anticipate: What are the possible disadvantages?

Anticipating Objections

If you know the drawbacks or pitfalls of what you are proposing but can demonstrate ways to minimize or avoid them, include a discussion of possible negative features in your proposal. For instance, is it not likely that the competitive market will peak by the time your company enters it? Might your company risk entering the competition with too little, too late?

If you suspect that this possibility will occur to your readers, try to provide reasonable reassurance. For example, suggest that to avoid entering the market after it has peaked, your company should manufacture only those hardware components which it is already equipped to turn out and should omit from this new product line any items that necessitate extensive re-tooling and retraining.

Step 4

By the time you get to this last step, you can see your best strategy. Your opening summary will be that this is a good time for your company to re-enter the computer hardware market because (a) unlike the time it first attempted to enter the market, there is now a competitive climate; (b) computer products would be compatible with the company's present product mix; and (c) some of its major competitors have already re-entered this market. The rest of the proposal would then document your three main reasons, anticipating and answering any questions or objections whenever possible.

CHOOSING AND REVISING PROPOSAL FORMATS

An excellent way to sharpen your proposal-writing skills is to read and analyze proposals written by others. As you learn to recognize what works well and what does not work well in other people's writing, you can apply this knowledge to the way that you write.

The following are two proposals that were written for a technical writing class. For each, we will analyze the format by applying four criteria:

1. Does the proposal begin with a clear and thorough opening conclusion?
2. Does it address the major concerns from the intended readers' point of view?
3. Does it anticipate and answer possible questions and objections?
4. Do the headings, paragraph structure, language, etc. make the information easy to understand?

Example 1

Example 1 (Figure 11.2) was written to the department supervisor of the Metallurgy Department of a Lamp Manufacturing Company by an entry-level metallurgical engineer. In a writing workshop this proposal was analyzed using the four criteria just illustrated.

Criteria 1: A Thorough Opening Summary

The participants agreed that the opening was clear and was complete enough to create the context needed to understand the body of the proposal. It is immediately apparent, even to an outsider that, according to the proposer, moving the Buehler Mounting Press from Room 335 to Room 129 would increase efficient use of time and space in three ways. And we are told immediately what the three ways are.

PROPOSAL TO MOVE THE BUEHLER MOUNTING PRESS FROM ITS
PRESENT LOCATION

Better use of space and time could be gained by moving the mounting press from Room 335 to Room 129. Such a move would accomplish three things. It would eliminate the need for an air compressor by making feasible a tie-in with the building's high-pressure air; it would free needed space for expansion of the transmission electron microscope (TEM) sample preparation lab; and it would help consolidate the metallographic equipment.

At present, the press is being run by a portable air compressor. Because of the location of the press, this has several disadvantages. While the compressor runs only intermittently, it is very noisy, making conferences and telephone calls impossible during those periods. There is enough vibration created when it is running that the compressor very likely has adverse effects on results from the microhardness tester located in the next room, especially when light loads are being used. The compressor also takes up valuable floor space. Piped-in high-pressure air is available in some parts of the building. To extend it to the present location of the press would involve going in between floors and would probably take many weeks, according to K. Anderson. An extension, of the high-pressure line to a first-floor location would be relatively simple since the line already extends from the high-bay area through the machine shop on that floor, five rooms away from the proposed location of the press.

Another advantage to moving the press would be to free counter space which is needed for the TEM sample thinning equipment. The equipment is crowded into a corner where it has become difficult to work as equipment is added. Etching tanks must be jockeyed depending upon the stage of thinning and setup needed. It is potentially a safety hazard where strong chemicals and awkward equipment are handled in a restricted space. Use of the Mettler balance is also restricted by the present arrangement of the thinning equipment. Very little space would be given up in Room 129 with the addition of the mounting press if it were to replace the press that is now in that room. The press that would be replaced is outdated and seldom used.

The majority of the polishing equipment is located on the first floor, either in Room 129 or Room 143. It would make good sense to have the mounting press located in the same area, so that samples need not be carried up and down two flights of stairs. Restriction on accessibility would not be a problem for those who are now using the press, if it were to be moved.

Figure 11.2   Proposal: first draft.

Criteria 2: Addressing Major Concerns from the Readers' View

Here, too, the group agreed that the proposal addresses the problem from the supervisor's perspective. Obviously, this proposal was prompted by the inconvenience the writer and her colleagues were experiencing because of the press's location. Her attempt to get the location changed could have been expressed in the form of a complaint letter. Instead, the writer wisely decided to take a "You" attitude. Rather than protesting that her needs were not being met, she formulated a proposal that would meet her supervisor's need: to operate his department at peak efficiency.

Criteria 3: Anticipating and Answering Questions or Objections

This criteria gave the workshop some pause. Finally, they noticed that most of the proposal is about improving operating procedures in Room 335 by removing the press. Very little is said about the effect of adding the press to Room 129. Then, one participant observed that tacked on to the end of the third paragraph were two sentences which anticipated what would happen in Room 129 if the press were placed there.

Criteria 4: Style and Ease of Understanding

The observation of these two sentences called the group's attention to the physical organization of the proposal. While they felt that the proposal was easy to understand after careful reading, they began to realize that there were ways of making it easier to grasp quickly.

The group's revision (Figure 11.3) makes the following changes in format to aid the reader's immediate comprehension of the different parts of the proposal and of how they are related to the main point:

Opening Conclusion

- Unnecessary words are omitted.
- The three reasons are numbered.

Body

- Each of the three reasons for moving the press is discussed in a separate paragraph.
- Appropriate headings are given to each paragraph.
- Each of the three paragraphs is divided into two labeled parts: the "Problem" and the "Solution."

Closing

- Because the original version has no ending, a conclusion is added which summarizes the main points and incorporates the answer to objections regarding what might happen in Room 129 if the press were moved there.

The revised version (Figure 11.3) facilitates understanding by doing the sorting and labeling of the information that the readers would have to do for themselves in the original version.

PROPOSAL TO MOVE THE BUEHLER MOUNTING PRESS FROM ITS PRESENT LOCATION

Better use of space and time could be gained by moving the mounting press from Room 335 to Room 129. Such a move would (1) eliminate the need for an air compressor by making feasible a tie-in with the building's high-pressure air; (2) free needed space for expansion of the transmission electron microscopy (TEM) sample preparation lab; and (3) help consolidate the metallographic equipment.

1. Elimination of Air Compressor

Problem: At present, the press is being run by a portable air compressor. Because of the location of the press, this has several disadvantages. While the compressor runs only intermittently, it is very noisy, making conferences and telephone calls impossible during those periods. There is so much vibration created when it is running, that the compressor very likely has adverse affects on results from the microhardness tester located in the next room, especially when light loads are being used. The compressor also takes up valuable floor space.

Solution: Piped-in high-pressure air is available in some parts of the building. To extend it to the present location of the press would involve going in between floors and would probably take many weeks. However, an extension of the high-pressure line to a first-floor location would be relatively simple since the line already extends from the high-bay area through the machine shop on that floor, five rooms away from the proposed location of the press.

2. Freeing of Space for TEM

Problem: At present, TEM sample thinning equipment is crowded into a corner in Room 335 where it has become difficult to work as equipment is added. Etching tanks must be jockeyed depending upon the stage of thinning and set-up needed. Handling strong chemicals and awkward equipment in a restricted space creates a potential safety hazard. Use of the Mettler balance is also restricted by the present arrangement of the thinning equipment. Very little space would be given up in Room 129 with the addition of the mounting press if it were to replace the press which is now in that room. The press which would be replaced is outdated and seldom used.

Solution: If the mounting press were moved out of Room 335, the proper space for utilizing TEM equipment safely would be available.

3. Consolidation of Metallographic Equipment

Problem: At present, the majority of the polishing-equipment is located on the first floor, either in Room 129 or Room 143. Consequently, samples from the mounting press must be carried up and down two flights of stairs.

Solution: If the Buehler Mounting Press were also in Room 129, samples would be easily accessible at each step of operation.

Figure 11.3   Revised proposal.

The transfer of the Buehler Mounting Press from Room 335 to Room 129 would increase operating efficiency in both of these work areas. The press that is situated in Room 129 presently is out-of-date and seldom used. If it were replaced with the Buehler Press, more space would be made available in Room 335 without sacrificing useful space in Room 129. Operators in both rooms would have all the equipment they need in one work space, and the hazards of overcrowding would be eliminated.

Figure 11.3 *(Continued)*

Example 2

Example (Figure 11.4) was written by an undergraduate Communications major who worked part-time and summers at a machine company. The most striking and effective parts of this proposal are the diagrams, which give the reader a "before" and "after" look at the flow of information in the company as it is and as it might be. The proposer uses these diagrams to demonstrate his main point: that the company would benefit from the employment of an organizational analysis service.

Does the proposal meet the four criteria we have been applying?

1. There is a clear statement of the *opening conclusion* in the first two paragraphs.
2. The concerns of the Vice President of Operations are focused on—both his general concern with the communication aspect of Operations and his personal concern about being overwhelmed with too much input.
3. The reader's possible objection to cost may be answered in the final paragraph by the reference to aid from the parent company. However, the reader is left pondering what would happen if the machine company had to cover the entire cost and how, if at all, the cost might be paid back in tangible improvements.
4. The format of the proposal is easy to follow, and headings are used effectively.

If you apply the four criteria used to evaluate these samples to your own proposals, your writing will be convincing and effective.

Mr. Joseph Hart
Vice President of Operations
Crown Machine Division of CBA Industries
147 Drake Avenue
Cleveland, Ohio

Dear Mr. Hart:

I am presenting the following proposal in response to the production meeting (November 16) in which you expressed your concern over the "inefficient handling of operations" and the "lack of cooperation" at Crown.

Proposal

Because these problems are a direct result of a "severe" breakdown in communications within the company, I propose that Crown Machine Division of CBA Industries engage the help of an outside organizational analysis service to establish an effective communications system within our organization's infrastructure.

Problem Description

Having been employed by Crown for four years on an alternating full-time/part-time basis, I have had the opportunity to assume the responsibilities of a number of positions in various departments (Figure 1). This experience has allowed me to witness and compare inter-departmental communication and interaction among personnel within each department. Based on these observations, I conclude that Crown Machine transmits information ineffectively in the following situations: face-to-face discussions, memos, telephone calls, reports, letters, and most other forms by which information is exchanged. Our company's problem seems to be inherent in its hierarchical infrastructure (Figure 2). Present communication directions and areas where communication is lacking are illustrated in Figure 3.

Departmental Employment Status

Summer (4 Months)	Winter (8 Months)
1984 Purchasing	1984 Documentation
1985 Service	1985 Production Assistance
1986 Expeditor (General)	1986 Expeditor (in-house)
1987 Subcontracting	1987 Subcontracting

Figure 1

Figure 11.4 Unsolicited proposal.

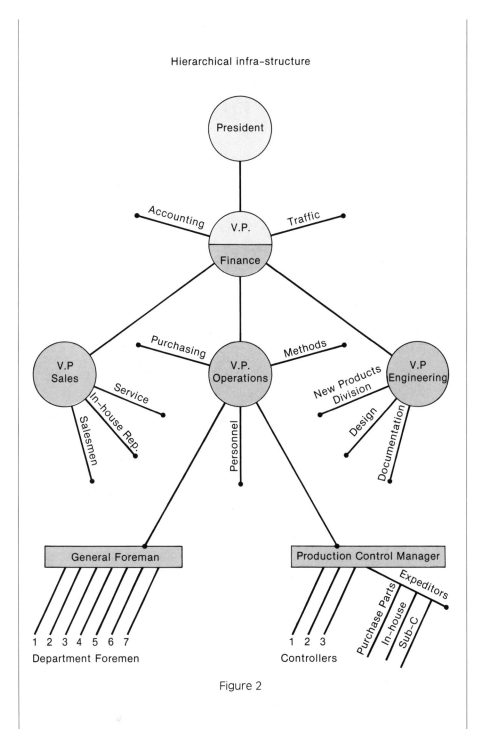

Figure 2

Figure 11.4 (*Continued*)

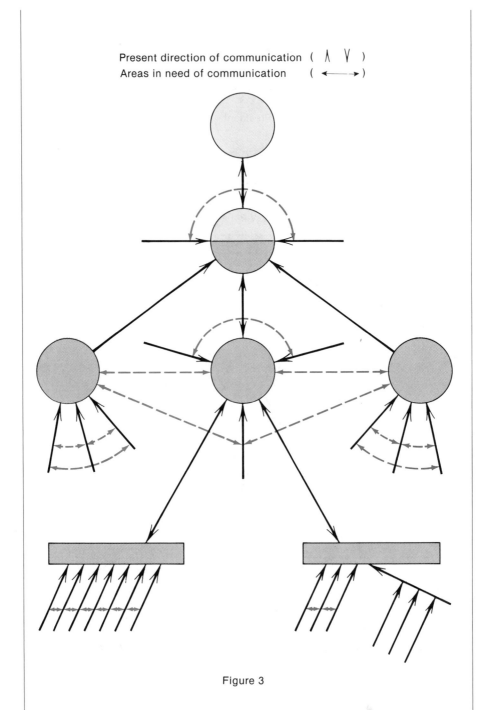

Present direction of communication (∧ ∨)
Areas in need of communication (←——→)

Figure 3

Figure 11.4 (*Continued*)

Suggested solution to present network of communications

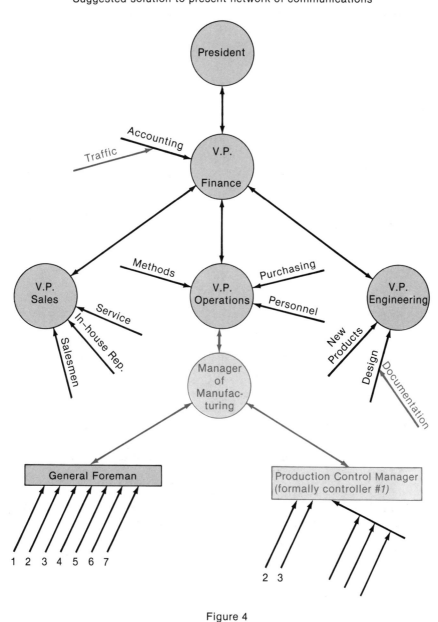

Figure 4

Figure 11.4 (*Continued*)

There are two major blocks in the system:

1. Some areas are experiencing communication overload.
2. There is a general lack of inter-departmental communication.

Because it is difficult for persons working within such a structure to view it objectively, we would benefit from an outside consultation.

Procedure

The two approaches an Organizational Analysis Service might use to determine communication effectiveness are (1) experimention and (2) description. The experimental approach involves setting up some communication structure and asking people to work on some problem using the communication channels established by the researcher. In this way, different types of communication patterns can be compared.

The descriptive approach centers on an investigator who tries to find out who talks to whom, about what, and how much in some ongoing group or organization. In this way, a picture or description of the process can be developed and certain problems uncovered.

Since approach 2 seems better suited to our company's problem, I compiled some of the data that this approach would require and turned it over to an organizational analysis consultant, along with a copy of the infrastructure diagram. The changes he recommended in our company's communications network are illustrated in Figure 4.

Basically, we have a centralized communications structure, which does not work as well as a decentralized structure in the handling of complex problem-solving because

1. The leader or person in the middle becomes overloaded. He has too much information to be able to function effectively.
2. The centralized structure does not allow the other members to have as substantial an input as they would in a decentralized structure.
3. Specific members are often isolated from the exchange, and satisfaction with the group seems to decrease.

To resolve these problems, the consultant suggests the creation of a Manager of Manufacturing who would transfer some of the burden of communicating with the General Foreman and the Production Control Division away from the Vice President of Operations. This is just one example of how our situation might be improved.

Figure 11.4   (*Continued*)

220

Cost

The cost of an outside Organization Analysis Service is considerable—no matter what approach is used in analysis. The length (in time) of the project, the number of researchers used, and the amount of research materials consumed are the determining factors. Below are actual costs for the communication network evaluation and problem resolution for TEREX Machine, a company with a hierarchical infrastructure very similar to ours:

Approach 1   Length 3 months
                Number of researchers 2 × $5,000
                Consumables $3,000

Equation   3 × (2 × $5,000) + $3,000 =
              $33,000 (subtotal)

Approach 2   Length 6 weeks
                Number of researchers 2 × $2,000
                Consumables ($1,400)

Equation   1.5 × (2 × $2,000) + $1,400 =
              $ 7,400 (2nd subtotal)
            + $33,000 (1st subtotal)
              $40,000 Total

Feasibility

In-House solutions to our internal communciation problems have proved to be only temporary at best. An outside service will have the advantage of being able to maintain critical objectivity, and is therefore our best option.

In addition, since our parent company stresses organizational efficiency and has subsidized retraining efforts in the past, chances are that they would reward our initiative this time by absorbing part of the cost of an outside consultant.

Thank you for your consideration.

Sincerely,

Arnold Martin

Figure 11.4   (*Continued*)

EXERCISES

1. With the aid of your school or office librarian, find a proposal written by a professional in your field of study. Read and evaluate the proposal according to the four questions listed on page 211.

2. Using the planning and writing techniques explained in this chapter, write a proposal to someone at your university or your office recommending a change in procedures or a solution to an operating problem. In group workshop, ask your fellow students to evaluate your proposal according to the four criteria listed on page 211.

CHAPTER 12

Writing Feasibility and Progress Reports

In the course of setting up, carrying through, and completing a project, the persons involved will often find it necessary to write letters, memos and reports well before the final report on the project is due. Some of these internal communications, such as trip reports and meeting minutes, have already been discussed in previous chapters. In this chapter, we will consider two of the most frequent types of writing that you are likely to do once the proposal of a project is being seriously considered or has been provisionally approved: the feasibility report and the progress report.

The Feasibility Report
A project is feasible if the time, equipment, and manpower necessary to complete it are cost efficient. To demonstrate feasibility,

- begin with an opening conclusion stating whether and why, the project is, is not, or might be feasible;
- clarify the methods and limits of your investigation;
- present the evidence in descending order of importance;
- close by re-stating your opening conclusion.

The Progress Report
The purposes of progress reports are

- to keep everyone involved in the project informed of what each one is doing,
- to make schedule adjustments possible throughout so that the final deadline can be met or renegotiated for verifiable reasons,
- to make supervisors aware of what you are accomplishing.

Structure your progress report either by periods of time or by numbers or kinds of tasks. With either format

- begin with an opening summary of what has been accomplished and what remains to be accomplished;
- include actual or anticipated delays and what can be done about them in the opening summary;
- present supporting evidence;
- conclude by re-stating the main points.

THE FEASIBILITY REPORT

Demonstrating that a project is feasible means demonstrating that it is practical—that the company has or can get the money, manpower, and equipment necessary, and that the project can be completed in a reasonable amount of time to make it worthwhile. As you can see, demonstrating feasibility is often part of the initial proposal itself. To convince your reader, your proposal must show the feasibility of your idea. Just as frequently, however, a proposal is made for a procedure that will test the feasibility of a project before it is undertaken or before it is implemented full-scale. For instance, a recent proposal made to the United States government outlines a program "to demonstrate the commercial viability of synthetic fuels." Another need for a feasibility report occurs when a supervisor, responding to an administrative request or an idea from the company suggestion box, asks a subordinate to test the feasibility of the idea and write a report. Depending on the subject matter, preparing a feasibility study may require library research, live tests, simulated tests, interviews, personal opinion, or more than one of these in combination.

To a certain extent, feasibility is in the eyes of the beholder. In other words, as in all the technical writing you do, the needs and concerns of your readers will strongly influence your approach. If you are filing an environmental impact statement with the Environmental Protection Agency, for example, the feasibility of your project will be determined on the basis of whether it has a negative impact on natural resources or human beings and, if so, to what extent these detrimental effects can be prevented or minimized. If you are a management consultant doing a feasibility report on whether a particular type of supermarket chain in the New England area should expand to the Midwest, then you will demonstrate viability based on whether a demographic profile of the area shows a sufficient number of residents whose profiles match those of the chain's present clientele.

Planning a Feasibility Report

Suppose that you have completed gathering the information necessary to determine the feasibility of a project. You have made use of hands-on testing, the reports of others, or a combination of both. You have a clear understanding of the organizational context: you know who the feasibility study is being prepared for and the purpose it will serve. In other words, you have before you the two lists or outlines discussed in Chapter 4—the content description and the reader description. In this case, your content description, tells you either that your information demonstrates feasibility, that it demonstrates lack of feasibility, or that it is inconclusive. This is the information that your readers want *first*. Then, they want to know any useful background information; a detailed description of what procedures you used—what you did and, where relevant, what you did not do—to determine feasibility; and, if requested, what recommendations you have for future action.

In the sample feasibility report in Figure 12.1, *paragraph 1* announces immediately that the production of a bridgeless lamp is feasible (sentence 1); reviews

Figure 12.1 Feasibility report.

MC/DC LAMP BRIDGE REMOVAL FEASIBILITY REPORT

Production of a bridgeless MD/DC lamp is a feasible project. A bridge (sometimes referred to as a bead or cane) is used to hold the lamp in place during the pinching operation. Eliminating the bridge and bridging operation will save $55K in material costs and $25K in labor costs for a total annual savings of $80K. Testing has shown no increase in shrinkage due to filament distortion. A minimum of retooling is necessary, and implementation should be possible during the first quarter of the next fiscal year.[1]

The Bridged Procedure Versus the Proposed Bridgeless Procedure

The mount assembly procedure, as required by the current design, is to load the spudded, foilless mount into the birdge paddle, place the quartz bridge into the bridge slot in the paddle, and insert the paddle into the bridging machine. Upon completion of the heating and pressing cycle, the paddle is removed from the bridging machine, and the bridged mount is then removed from the paddle. The bridged mount is then placed in a welding fixture, and the foils are welded to the spuds. The mount is now ready for pinching.[2]

The mount assembly procedure for the new design would eliminate the bridging operation. The foil welding operation would be performed immediately after the spudding of the coil. New welding fixtures will be needed to hold the bridgeless mounts in place and properly locate the foils on the spuds. However, since a program is already in progress to replace the current textile fixtures with stainless steel fixtures, no additional expenses will be incurred.[3]

The only potential disadvantage of eliminating the bridge is that increased shrinkage might result from filament distortion. To predict the likelihood of significant shrinkage, a test was run using a bridgeless mount at pinching. From a run of 500 mounts, only two show signs of filament distortion, and these were attributable to poor welding from using a mock-up fixture. With stainless-steel fixtures, the accuracy of the weld should be no problem.[4]

Based upon the favorable results of testing and the high cost savings projected, production of MC/DC lamps with bridgeless mounts is deemed a feasible project.[5]

how the bridge is used (sentence 2); recalls the original reason for considering the elimination of the bridge (sentence 3); and, finally, summarizes the reasons for considering the project feasible (sentences 4 and 5).

The *second and third paragraphs* use a comparison format to explain the difference between a mount assembly procedure with a bridge (the present method), and a similar procedure without the bridge (the proposed method).

Part of the *third* and all of the *fourth paragraphs* discuss the possible disadvantages of the new process and eliminate them as reasons for rejecting the new process.

The *closing paragraph* (5) reiterates the reasons for feasibility and follows naturally from the evidence presented in paragraphs 2 through 4.

Demonstrating That a Project Is *Not* Feasible

Just as often a feasibility report demonstrates that a desired change in procedure or a new project is not feasible. Although negative conclusions are disappointing, these feasibility reports are just as useful as those with a positive conclusion. Being able to identify a project as untenable in the early stages saves needless commitment of time, equipment and manpower and enables the company to employ its resources in search of alternative solutions which may prove more feasible.

Reports which document a procedure as unfeasible are also an important addition to company records. If an engineer has a particular problem-solving technique in mind, he or she can check the files to see if anything similar has been tried before and under what circumstances. Unnecessary duplication of effort is avoided, and, sometimes, a technique that was not feasible under one set of circumstances is discovered to be viable in a different application.

Example: The Negative Conclusion

Figure 12.2 is a feasibility study that resulted in a negative conclusion. The purpose of this study was to determine the desirability of implementing one of many possible solutions to an ongoing production problem. Many feasibility studies are generated by a quality control engineer's discovery of what is causing equipment failure or defective production above the acceptable error margin. If there is no obvious or standard procedure to eliminate the problem, the Design Department or the Research and Development Department is handed the task of proposing possible solutions. Each of these may be theoretically sound, but must be tested for practicality. Many prove unfeasible in actual operation, but through trial and error a feasible solution or the best of the lot is usually identified.

In the feasibility report on using thermal techniques to evaluate inside weld flash trim (Figure 12.2), the investigator demonstrates that this solution is not feasible because tests show that there is no significant correlation between temperature changes and variations in weld thickness. *Paragraph 1* summarizes the purpose of the study, the test performed, the result, and the recommendation.

FEASIBILITY OF USING THERMAL TECHNIQUES FOR ON-LINE
EVALUATION OF INSIDE WELD FLASH TRIM AT ABC PIPE MILL

The purpose of this study was to determine the feasibility of using thermal
monitoring techniques at the welder site to detect improperly trimmed inside
weld flash. A two-week test using an infrared thermal monitor demonstrated that
there is no significant correlation between weld flash trim variations and temperature changes. Consequently, thermal monitoring techniques are not a feasible replacement for present inspection procedures.[1]

Background

At present, the procedure for evaluation of inside flash trim calls for visual
inspection on a rack approximately 250 feet from the welding site. As a result, it
is sometimes necessary to reject as much as 250 feet of defective piping.[2]

If improperly trimmed weld flash could be detected nearer the weld site, less
footage of defective piping would be produced and significant savings in man-
hours and matrials could be achieved. A proposed solution is monitoring by temperature measurement at the weld site.[3]

Method

For a two-week period weld zone temperature measurements were made down-
stream of the X-12 station. A thermovision profiler and a Vanzetti infrared thermal monitor were used at the welder station to record temperatures at the weld
zone of the pipe.[4]

Results

The recorded temperatures varied as much as 100°F from front to rear of a single coil. At no time during the monitoring did the inside flash trim tool fail to
trim completely. However, during this period, trim variations sufficient to warrant trim tool changes did occur. These variables did not correlate with the temperature variations. The only detectable correlation was obtained when the
welder operator made slight variations in the mill speed and/or welding current.
These changes resulted in slight temperature changes (to 10°F).[5]

Recommendations

Due to the lack of correlation between weld flash trim variations and temperatures changes, this investigation should be discontinued.[6]

Figure 12.2   Feasibility report with a negative conclusion.

Paragraphs 2 and 3 state the problem, the general kind of solution that is
needed, and the specific solution that this report concerns.

Paragraph 4 describes the method used to test the proposed solution's
effectiveness.

Paragraph 5 describes the results observed.

Paragraph 6 recommends discontinuation of the study based on the lack of feasibility demonstrated in paragraph 5.

Either simultaneously or in succession, a number of other possible procedures for reducing the production of defective piping will be tested for feasibility, until one proves feasible or until it becomes clear that no more feasible method for reducing error than the one already in use is available.

THE PROGRESS REPORT

When you propose or are assigned to a project that will be continued over a period of time, the work is usually divided into a given number of phases, steps, and/or tasks. In order for the entire project to be completed by the deadline indicated—in 12 weeks, or by the third quarter of 1987—each stage must be completed on schedule or, when necessary, must be reevaluated and rescheduled so that despite unexpected setbacks, the final deadline can still be met. In the event that delays cannot be compensated for, a new deadline can be negotiated well in advance at minimum inconvenience to everyone concerned.

Progress reports, therefore, *have two major purposes*:

1. To keep everyone involved in the project informed on what each one is doing.
2. To make schedule adjustments and compensations possible throughout the project so that the final deadline can either be met as promised or renegotiated for legitimate, verifiable reasons.

A progress report, whether supplied in response to a request or on your own initiative, is also an excellent device for keeping your superiors aware of how much you are accomplishing.

Particularly in a project of one or more years' duration, progress reports are reference guides and provide much of the material used to prepare the final report when the entire project has been completed. Furthermore, if a project is discontinued for any length of time and then taken up again, participants can use the progress reports to determine quickly where their predecessors left off and where they need to begin.

Although it may seem like an imposition to have to stop work on an exciting project every month or quarter to fill out a form describing what you are doing, in actuality the progress report has several advantages for your company, for your co-workers and, even, for you. Furthermore, once you master the progress report technique, it becomes an opportunity to clarify your thoughts about your work periodically and get an overview of how the work you did in the last two weeks relates to the goal of the entire project.

Planning a Progress Report

If your company does not have a standard progress report form which requires filling in the blanks, then you have several possible formats from which to choose. Which one you select should be determined on the basis of which format demonstrates your progress most favorably without sacrificing accuracy.

You can structure your progress reports around periods of time, or around number or kinds of tasks.[1] In either case, your progress report should begin with an opening summary of either what has been accomplished in a given time period and what remains to be accomplished in subsequent time periods, or else of which tasks have been fully or partially completed and which remain to be completed. If you are behind schedule or anticipate problems with keeping on schedule, particularly for reasons beyond your control (e.g., undelivered parts, sick or uncooperative personnel, unexpected demands on your time), these significant delays should be mentioned in the summary. All the highlights of the progress report should be summarized in the opening, regardless of which format you are using.

Organizing by Periods of Time

Here, in outline form, are two variations on progress reports organized by periods of time:

A. Opening Summary
B. Phase I (3 weeks)—tentative selection of sites
    First week
       Work completed
    Second week
       Work completed
    Third week
       Work completed
C. Phase II (2 weeks)—comparison of selected sites
    First week
       Work completed
    Second week
       Work completed
D. Phase III (2 weeks)—bidding and negotiating on site selected
    First week
       Work to be completed
    Second week
       Work to be completed

A. Opening summary
B. Month 1
    First week
       Work completed
       Work remaining
    Second week
       Work completed
       Work remaining

[1]See Houp and Pearsall, *Reporting Technical Information,* 5th ed. (New York: 1984), Chapter 16.

Third week
  Work completed
  Work remaining
Fourth week
  Work completed
  Work remaining
Fifth week
  Work completed
  Work remaining
C. Month 2
  First week
    Work completed
    Work remaining
  Second week
    Work completed
    Work remaining
  Third week
    Work completed
    Work remaining
  Fourth week
    Work completed
    Work remaining

Organizing by Numbers or Kinds of Tasks

If you choose to organize by task, you reverse the procedure used in time-based organization. In other words, the time periods are subordinated to the tasks completed or remaining instead of the other way around. For example:

A. Summary of tasks completed and tasks worked on

B. Task 1—assembly of necessary equipment and personnel
      Work completed
        Week 1
        Week 2
        Week 3
        Week 4
      Work remaining

C. Task 2—training of personnel in use of new equipment
      Work completed
        Week 1
        Week 2
      Work remaining

Choosing Between the Two Report Formats

If you expect to make a lot of progress on some tasks and very little on others during each report period, then a time format is better. If you have several tasks, each of which will show only a little progress each time, then a task format is the more appropriate choice. It depends on whether you wish to stress how many tasks you have accomplished in a given time period or how much progress you have made on one or two large tasks in a short period of time. As in resumé preparation, you do not want to distort the information; you want to package it in the most attractive and efficient way possible.

Example: Using the Time Structure

Figure 12.3 is a progress report that subordinates task to time. After the introductory section, which reviews the purpose and background of the project, the body of the report is divided into three time categories:

- Past Work
- Present Work
- Future Work.

Under each time period, tasks are listed and described. For example:

Past Work
- Implementing Error Rate Detector
- Measuring Error Rates
- Determining Sources of Error.

Present Work
- Optimizing capacity
- Profiling time of day vs. error rate.

Future Work
- Determining if C is independent of transmission frequency
- Completing profile study for various frequencies
- Making final recommendations.

Example: Using the Task Structure

The writer of the progress report in Figure 12.4 organized her first four-week review around the two major tasks of her research project:

Introduction
Task I—infection of synovial and cartilage cultures with mycoplasma
- Purpose
- Procedure
- Work completed
- Work remaining

Task II—Injection of mycoplasms into lapine joints; culture of joint cells
- Purpose
- Procedure

MID-SEMESTER PROGRESS REPORT ON ERROR REDUCTION IN AN EXISTING DIGITAL TRANSMISSION MEDIUM

Introduction

This is a mid-semester report of the progress made toward determining and reducing error rates in an existing digital transmission medium. This report covers the period between January 15 and April 9, 1987. As of April 9 work is going as scheduled, with approximately 85 percent of the tasks accomplished. The expected completion date is April 23, 1987, and no problems are anticipated in meeting this deadline.

Purpose

A ring of uni-directional, twisted-pair transmission lines has been installed connecting four interface circuits on the fourth, sixth, and second floor of the Glennan building at Case Western Reserve University (CWRU). Before this digital transmission medium is used in practical data acquisition, an investigation into existing error rates must be made following this general procedure:

1. Design of a simple one-way transmission scheme to detect the rate at which erroneous transmissions occur;

2. Determination of whether the existing medium is indeed "noisy" (i.e., has high error rates);

3. If the line is noisy, isolation of the causes of noise and redesign where necessary to reduce rates to an acceptable minimum.

This report covers present progress in executing these tasks.

Background

The Mechanical and Aerospace Engineering Department at CWRU will soon put a data acquisition network into operation. This network will consist of up to eight distinct sites located throughout the Glennan building. Proper operation requires communication between each site in the form of serial data "messages". The medium used to accommodate these messages is a twisted-pair transmission line set up in a ring configuration. The transmission line is interfaced at each site by a circuit which receives the serial data and then routes it back onto the site hardware. A critical requirement of this transmission medium is that it be relatively free of errors. An acceptable error rate is 1 error per 10 billion transmission of bits.

Up until January 15, this medium had not been probed extensively to determine existing error rates. Furthermore, while it was used during routine testing of the digital circuits comprising each hardware site, the transmission lines generated their own "messages" when none were intended to be sent. Since proper operating characteristics of the medium are crucial to good performance of the network, this behavior led to the decision to probe thoroughly and redesign (where necessary) the transmission medium and to document the results for future reference. This report describes the steps taken already toward accomplishing these tasks, what steps are being taken presently, and what remains to be done before final recommendations can be made.

Figure 12.3 Progress report: time structure.

Implementing Digital Error Rate Detector: An error rate detector has been implemented using digital circuitry which reliably outputs the number of errors/65, 535 bits of transmission data as a four-digit hexadecimal number.

A scheme designed to detect the extent of error likely to occur in actual use must closely approximate actual operating conditions. To ensure that this will be the case, the author implemented a hardware detector which:

1. Generates as data a repeating 65,535 bit sequence proven mathematically to be "worst-case";

2. Sends this data stream through the medium at actual operating speeds.

"Worst-case" here means a sequence of bits which approximates a random sequence; this is necessary to allow for the possibility that for some reason one particular sequence pattern is likely to be more error prone than others. Also, since the transmission line is inherently less reliable as speed increases, sending data at operating speeds is critical in revealing the actual extent of error.

The implemented detector is designed to run in conjunction with an 8080 microcomputer, which reads the current error rate and signals the detector to sense further rates. The two have been operating well together along with a CRT (Cathode Ray Tube) terminal to give a visual readout of the error rates with time. A completely documented design will be included in the final report.

Measuring Error Rates

Using the error detector, the author was then able to "probe" the transmission medium for existing error rates. Transmitting at 1.25 Mbits/sec, an error rate of $7FFF_{16}$ (50 percent) was measured for the four sites connected together. For a two-site configuration (both sites located on the second floor and separated by about 20 feet of cable), the error rates were consistently zero.

Determining Sources of Error: The highly contrasting error rates found for these two configurations provided the first clue determining the possible sources of error. What followed was a systematic and largely empirical investigation intended to isolate the causes and implement the necessary design changes. The following steps were taken:

1. Between the two sites on the second floor, a number of connecting wires had been made to allow for a third site to be installed within a PDP-11-40 minicomputer housing. The author disconnected the confusing web of wires and in the process found that some twisted-pair lengths had been erroneously connected so that the number of twists per foot was 1.5 rather than 9. Since the number of twists per foot directly relates to the line's noise immunity, this was one indirect source of error which the author eliminated by bypassing it.

2. With the two sites' connections "cleaned up" in this manner, the author measured a 200-foot length of cable and electrically inserted it between the two sites to determine whether cable length might be a factor (about 100 feet of cable is presently the maximum length between any two sites). The resulting error rates increased and stabilized around 2000_{16} (12.5 percent error). Thus, apparently the interface circuitry was not configured to properly handle cable of this length.

Figure 12.3 (*Continued*)

3. The next step involved a close examination of the interface circuitry for probable design flaws. Since very little published information was available regarding the interface devices used or their application, the author called the device manufacturer, Signetics Corporation, to briefly discuss correct application use. As a result of this discussion the author found that the devices were being used incorrectly.

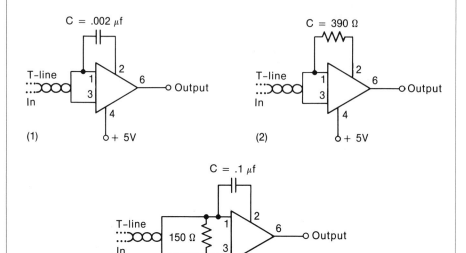

Figure 1 Recommended (1), previous (2), and subsequent (3) actual design for line receiver interface (Part Number: DS820).

As Figure 1 shows, the manufacturer recommends that a capacitor be placed between pins 1 and 2 of the device to match the capacitive characteristics of the transmission line (this matching is critical to enhance noise immunity). The author removed the incorrectly connected 390 ohm resistor and inserted a .1-microfarad capacitor. Also, to improve the shape of the transmitted signal, he inserted a 150-ohm resistor between the two ends of the transmission line itself.

After the design modifications shown in Figure 1 were made, the two sites separated by 200 feet of cable introduced zero errors consistently. When all four sites were similarly modified, the error rate for four sites also consistently stayed at zero.

PRESENT WORK

Optimizing Capacity

Although a value of $C = .1$ microfarad eliminated errors, the question remained whether or not this was an optimum value for the cable length of 200 feet. After comparing the output signal waveforms for values of C equal to .474, .1, .05, .04, and .01 microfarad, the author found that no difference in signal shape existed which might affect the rate of error. Subsequent error rate readings using these five values verified this fact (i.e., error rates stayed at zero).

Figure 12.3 (*Continued*)

Profiling Time of Day Versus Error Rate

A study to determine whether or not the time of day has any effect on the error rates has been halfway completed. Software for the 8080 microcomputer to communicate with the PDP 11 has been written and tested, along with a flow chart of the type of communication which will occur in the reverse direction.

This study will indicate whether periods of peak electrical usage within Glennan building have any adverse effects on transmission accuracy. In the same way that a power tool can distort a television transmission, digital data transmission on these lines can be affected. The PDP-11/40 will record and average error rates for the first 10 minutes of every hour throughout the day.

FUTURE WORK

1. Determining if C is independent of transmission frequency;
2. Completing profile study for various frequencies to determine upper limit on transmission speed;
3. Making final recommendations.

Task 1 will be delayed until an order for several crystal oscillators has arrived. Task 2 is in the process of completion; however, the frequency limit study also awaits the order. Task 3 will be performed after 1 and 2 and will represent sufficient documentation to allow for future reference in proper installation of more sites.

Figure 12.3 *(Continued)*

NIH-500114
October 13, 1986

Dr. Garrett McCory
Executive Chairman
NIH Board of Directors
National Institute of Health
Washington, D.C. 20036

Dear Dr. McCory:

Progress has been made on Phase I of the research project funded by Grant NIH-500114 in the first four-week period of the project. This grant has been allocated for the purpose of determining a possible relationship between mycoplasma infection of joint tissues and the development of rheumatoid arthritis. As required, this report summarizes experimental results for the first four-week review period (September 16 through October 14) and outlines future experiments.

Figure 12.4 Progress report: task structure.

In order to link mycoplasma infection of joints with the development of rheumatoid arthritis, it must be demonstrated that mycoplasma infection of appropriate joint tissues results in the production and secretion of levels of collagenase significantly higher than control levels. It must also be shown that increased levels of collagenase result in damage to joint material. Potential sources of collagenase production in rheumatoid arthritis are synovial and cartilage cells. Primary cultures, passage two, of rabbit synovial cells are being provided by Dr. L. Gross, Dept. of Rheumatology, University Hospital. Primary cultures, passage two, of rabbit cartilage cells are being provided by Dr. C. Mendez, Dept. of Medicine, University Hospital. Verification of mycoplasma infection of synovial and cartilage cultures will be made by Dr. E. Stanhope.

This research project has been divided into two phases:

Task 1: Infection of synovial and cartilage cultures with various strains of mycoplasma

Task 2: Injection of mycoplasmas into lapine joints; subsequent culture of the joint cells after infection

WORK STATEMENT (TASK I)

Purpose

The purpose of this section is to induce the production and secretion of collagenase in synovial and cartilage cultures following infection with various mycoplasma strains.

Procedure

Duplicate cultures of synovial and cartilage cells are infected with *M. orales* as well as *M. fermantans,* and *M. arthritidis.* Control cultures are not infected. Fresh medium (15 ml.) is added to the cultures and left for 48 hours. After this time period, the medium is removed and totally fresh medium is added to the cells. Similar collections are made every two days thereafter. The collagenase activity is measured in each 15-ml fraction by activation of the crude medium with p-aminophenylmercuric acetate, followed by an assay which measures the release of radioactive peptides from gels of native collagen. The rate of collagen degradation for each collection period is measured and plotted.

Work Completed

The first set of synovial and cartilage cultures were infected with *M. orales.* Assays were done on fractions collected at specific intervals between one-half and nine days following the addition of the first aliquot of medium. The results are shown in Figures 1 and 2. Data is plotted as milligrams of collagen degraded/hr/dish versus duration of culture. Collagenase activity begins to accumulate in synovial culture medium at about day 2, peaks at day 4, then declines slowly. Synovial control cultures showed no collagenase activity. Cartilage cultures showed only slight collagenase activity throughout the collection period. Cartilage controls showed no activity. These cultures have been sent to California to determine if they are infected. The results have not been received.

Work Remaining

The same procedure as listed above will be carried out for cultures infected with *M. fermantans,* and *M. arthritidis.* All experiments will be performed in triplicate.

Figure 12.4   *(Continued)*

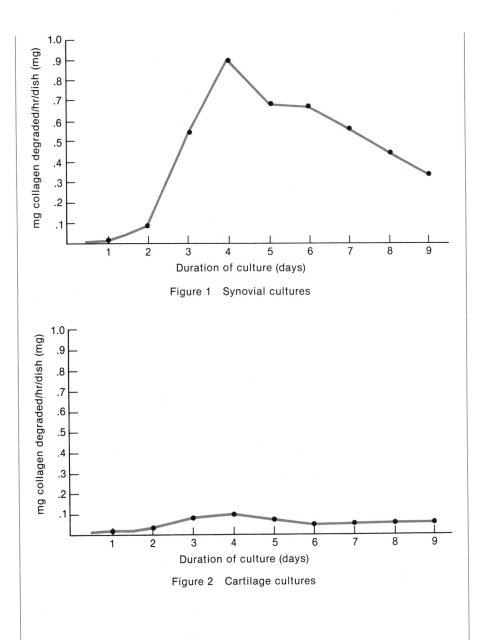

Figure 1   Synovial cultures

Figure 2   Cartilage cultures

WORK STATEMENT (TASK II)

Purpose

The purpose of this section of the project is to elicit the symptoms of rheumatoid arthritis in an animal model following the injection of mycoplasmas into joints. After infection, pieces of joint tissue will be removed and cultured. The production and secretion of collagenase from these cells will be measured.

Figure 12.4   (*Continued*)

237

Procedure

A certain amount of mycoplasma-containing medium will be injected into lapine joints. After two weeks, joint tissue will be removed and cultured. Collagenase activity will be measured in the same way described in Phase I procedure.

Work Remaining

This experiment has not yet been initiated. I am planning to begin this phase in six weeks. The induction of rheumatoid arthritis following the injection of the different mycoplasma strains into joints will be studied. The same procedure as listed in Phase I will be used to study the collagenase activity in joint tissues.

OVERALL APPRAISAL

All materials and facilities needed to conduct this project have been procured. The first task of the project seems to be proceeding as planned. Minimal data have been collected; however, the results look hopeful in terms of confirming the expectations proposed by my hypothesis.

Figure 12.4 *(Continued)*

March 3, 1987

SUBJECT: PROGRESS REPORT ON BALLAST CATALOGUE UPDATING

This report brings you up to date on the progress I have made on the revision of the existing ballast catalogue. It appears that there is still some substantial information gathering to be done to prepare for printing a new catalog, but several issues have been resolved.

Tasks Completed

1. The current catalogue was compared to the internal product master. It was found that 115 types listed on the master were not found in the catalogue. During the recent trip to Pleasantville, I was able to identify partial descriptions for all but eight of these types.
2. The current catalogue contained six product types not included on the master. In Pleasantville, it was discovered that four types now have different product codes; the other two types have been delisted.
3. The current catalogue contains 14 5W-type ballasts. It was learned that these have all been delisted and can be eliminated from the catalogue.
4. Technical data available in advertising was secured as a backup description reference.

Tasks Remaining

1. Confer with Jim Phillips in Pleasantville to provide full data required for all types.
2. Study most effective means of keeping an updated reference for the ballast product line, such as a stored file on computer or word processor.
3. Receive monthly master to keep up to date.
4. Determine effects of delistments.

Figure 12.5 Progress report: task structure 2.

- Work completed
- Work remaining

Overall appraisal

Figure 12.5, unlike the previous two examples, was generated by the writer herself rather than in compliance with a request from a supervisor or a grant stipulation. The writer uses this task-formulated progress report to inform her supervisor of two facts: (1) she understands the full scope of the project she has been assigned, and (2) she is making progress in getting the job done.

Often, a manager with a large staff will not be able to keep track of all the assignments given out at all times. Frequently, managers will not request a status report unless a sudden need arises or until they receive requests for information on the project from their managers. Therefore, it is a good idea to write or be ready to write a progress report on anything you have been assigned, even if it is not initially required. That way, you can handle unexpected requests for reports, or, if no request is made, you can profer a progress report on your own initiative.

Both feasibility and progress reports are useful tools of every profession. They can benefit both you and the people you work for by saving time, improving efficiency, and providing a convenient flow of communication among staff members whose goals are often interdependent.

EXERCISES

1. Write a *feasibility report* on some project you have considered undertaking at your university, office, or home. For example, have you been planning to rebuild a car, re-upholster your old furniture instead of buying new pieces, run for election to student government, or update your filing system? Evaluate the feasibility of your intended project and, if possible, exchange reports with at least one other person in your class to get reader feedback from different sources.

2. Write a *progress report* on a current activity: a term paper, a do-it-yourself project, community volunteer work, your attempts to get into medical school, or to get a job. Outline the structure of your report to identify whether it is organized by time or by task. Exchange reports with classmates for helpful reader feedback.

Writing Long Technical Reports

To write a long report, apply the techniques you have been learning throughout this book. Sell your long report by convincing your reader that the essential tasks have been performed and that the information supports the conclusions (Chapters 1 through 3). Provide your reader with opening conclusions (Chapter 3): in a long report, one to give an overview of the main points, and a second to provide an outline of the report's structure.

Before you write, prepare a reader-description list and an information-description outline (Chapter 4). While you write, keep to the structure you have chosen; make smooth transitional statements; incorporate graphic aids where needed, and close by repeating the main points in your opening conclusions (Chapters 5 and 6).

Precede all detailed object or process descriptions with an overview of function or purpose (Chapter 10); anticipate possible reader objections or questions; and incorporate explanations or answers into your report (Chapter 11). Provide a schedule of progress for the project, and make any limitations of your study clear from the beginning (Chapter 12). After you write, proofread twice—once for sense, and once for mechanical correctness (Chapter 7).

THE SPECIAL FEATURES OF LONG REPORTS

A long technical report is any written communication of five pages or more which requires the coordination and presentation of complex technical data of different kinds from many different sources. Usually, it is written after a major project or a significant stage of a project has been completed. The writer's challenge is to combine different kinds of supporting documents such as graphs, tables, journal articles, interim reports, experimental data, and the comments of others, in order to produce a smoothly flowing report that makes the main points and supporting evidence easily accessible to the intended readers.

To demonstrate the integration of skills necessary to prepare long reports this chapter examines selected problems in three case studies:

1. A final laboratory report by a metallurgical engineer entitled "The Impact on Hot-Strip Mill-Rolling Resistance of Additions of Carbon, Aluminum, Molybdenum, and Silicon to Plain Carbon Steels" (10 pages)
2. A research report by an accountant entitled "Design and Implementation of Price Forecasting Models for the Red-Meat Industry" (40 pages)
3. A final environmental impact statement entitled "Alternate Fuels Demonstration Program" (2 volumes, 1162 pages)

CASE STUDY SAMPLES

Example 1: A Final Lab Report

Frequently, an engineer working in the Research and Development Division or the Quality Control Division of a manufacturing company is asked to investigate ways to improve the processing of a product. The assignment may be part of a routine improvement program; it may be in response to a new use for a product which requires altering its properties; or it may be initiated by a need to increase cost efficiency.

Whatever prompts the project, the engineer's job usually entails reading the available literature on the subject, performing hands-on or simulated tests, and comparing published results with his or her own observations in order to make a recommendation or provide the information on which a recommendation by management can be based.

The process of concern in the sample report is hot-strip mill rolling of plain carbon steel. The steel's resistance to rolling is a problem, and in recent literature four different substances have been suggested as possible additions to a plain carbon-base chemistry that will increase pliability: (1) carbon, (2) aluminum, (3) silicon, and (4) molybdenum.

The engineer/writer's tasks are to study previous regression analyses using each of these four substances; to perform his own regression analysis and compare the results; and to put his observations into a report.

Before Writing

Since he recognizes at the beginning that his report will require a comparison format (supplemented by descriptions), he can divide his subject into four subtopics, one for each of the four substances he will consider. Knowing the criteria for establishing significant resistance reduction and the possible variations in test procedures (e.g., thickness of steel, percentage of substance), he can set up his outline as a comparison frame (Figure 13.1) and fill it in as he proceeds with his reading and testing. When all his research is completed and he has used his subject-coded notes to fill in his comparison chart, the engineer can evaluate

	Carbon	Aluminum	Silicon	Molybdenum
Tests #	1 2 3 4 5 6 7	1 2 3 4 5 6 7	1 2 3 4 5 6 7	1 2 3 4 5 6 7
Number of Passes				
Percent of Reduction				
Temperature				
Chemistry Level				
Steel thickness				
Steel weight				
Other Variables				

Figure 13.1   Comparison frame.

the significance of his observations and determine an opening conclusion for the "Summary" section of his report.

In this case, the frame reveals statistically significant resistance reduction by two of the four substances considered. In the simulated tests performed by the engineer the effect of one of these two elements appeared to increase at each additional pass.

As You Write

The *report summary* will include the following:

Problem: Greater demand for products using hot strip mill rolled carbon steel has created a need to reduce this type of steel's resistance to rolling.

Methods: Using the guidelines of past literature and simulated lab tests, each of four elements was added to the plain carbon-steel formula to reduce resistance.

Observations: In both the literature and our own tests, carbon and aluminum did not alter resistance, while silicon and molybdenum caused statistically significant reduction.

Results: In the lab procedures the effects of silicon appeared to increase at each additional pass, a factor not indicated in previous literature.

Recommendations: More passes should be run under more varied conditions to verify the consistency of this effect.

The introduction to the body of the report will repeat the problem, method, observations, and results in brief but will also give an overview of procedures including limitations or special circumstances. The introduction prepares the reader to interpret the detailed descriptions of test conditions, operations, and results. For example:

Thirteen chemistries were melted in the laboratory (see Figure 1). Each element studied has four weight percent levels. Eight plates were rolled for the base chemistry and two for each varied chemistry. The replicate rollings (see Figure 2) allow an estimate of the experimental repeatability. Neither the chemistry sequence nor the laboratory rolling sequence correspond to common steel grades or mill/C rollings. The chemistry sequence was necessary to derive mathematically independent estimates of effects. The rolling simulates hot strip mill processing.

The body of the report will first summarize and then detail (1) the tests studied and (2) the tests performed. Because understanding the report necessitates "seeing" the results of the test, descriptions will focus on graphs recording the results observed (see Figure 13.2). Ideally, each graph should be on the page facing the page of the test which refers to it. If your company's procedure is to place all illustrations in an appendix, the report should be bound so that appendix pages are removable. Diagrams that must be looked at in conjunction with reading should be placed beside the relevant description by the writer or should be able to be moved there by the reader. Any necessary explanations not indi-

Figure 13.2   Rolling load simulation.

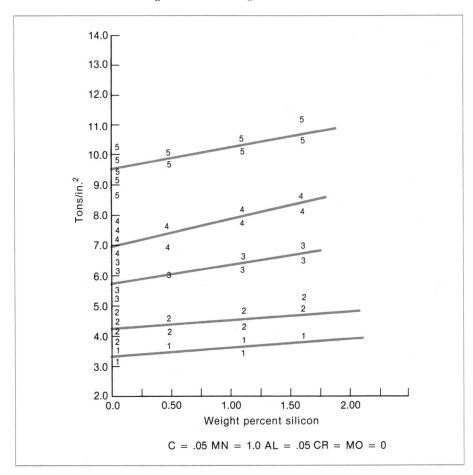

cated by labels on the diagram should be included at the bottom (see Figure 13.2).

After tests using each of the four elements have been summarized and described individually, the conclusion will repeat the main points stated in the summary and the introduction:

CONCLUSION

Both the previous literature and our own simulated passes indicate that of four possible additions to the basic formula of plain carbon steel, carbon and aluminum do not significantly decrease rolling resistance; silicon and molybdenum do demonstrate statistically significant decreases in resistance. On our own tests, although not in the literature, the silicon chemistry exhibited an increased effect with each pass. Further tests should be run to confirm the effects of silicon, and preparations should be made for a trial on-site run of the new steel.

After You Write

The engineer will reread his report twice: once for clarity and conciseness; once to check out consistency of information and appropriateness of language and format to audience.

Because he has been organizing his report from the moment he began his research, the engineer has an excellent grasp of his information and can respond with confidence to reader feedback.

Example 2: A Final Research Report

Often on their own initiative or in response to a request for an investigation, employees will discover a method to increase company profits. Long reports written to present such a discovery use the persuasion structure and are frequently the basis for a proposal. But they also require a description of the process, instructions in how to use it, and comparisons with old or alternate methods.

In this example an accountant employed by a meat-packing company has devised a price forecasting system to increase profits. She contends that because of short-term changes in the market prices of livestock and packaged beef, the company that predicts these changes accurately can increase its profits by purchasing livestock when it is cheapest and selling packaged meat when it is most expensive.

Her reader-description list shows that the top-management audience is likely to expect answers to the following questions:

1. What are the available models for price forecasting in our industry, and how do they compare?
2. Is there a model that would work for us? How does it work?
3. How expensive is it? Will the cost of maintaining accurate forecasts wipe out the profits?
4. What are the liabilities?

Because her decision-making readers are not economists, she will have to make them understand the principles of forecasting in order to sell them on it. Part of her report will be an instruction manual.

Using the technique recommended in Chapter 4 for organizing a long report as information is accumulated, the accountant already has identified the necessary sections of her report and has divided her information under each section. Reading the outline summary of her notes enables her to develop an opening conclusion (Figure 13.3) and a logical format for the supporting facts. As indicated in the table of contents (Figure 13.4), the report moves from general to specific: from how prices impact profits to how forecasting models predict prices, to a comparison of forecasting types, to a detailed description of an actual model for the company (with sample applications).

In the section on "Implications of Forecasting Models" (Figure 13.5), the writer begins with the general statement that exponential smoothing models work best during a volatile market and moving average models work best during periods of stagnation. The ideal system is set up to switch from one to the other at the right time. To demonstrate, the accountant details price changes on an average Monday through Friday, day by day. She refers to the relevant columns in Tables 4, 5, and 6 to verify her analyses.

Graphics

Because the writer refers to the same few tables and figures throughout her report, placing each adjacent to each page of text that refers to it would not facilitate her readers' task. Instead, she has placed all the illustrations in a pocket attached to the back cover of the report. That way the reader can lift out the appropriate chart each time it is referred to, follow along on it when necessary, and then place it back in the pocket until it is needed again.

Figure 13.3   Opening conclusion.

SUMMARY

A red-meat packer can operate more profitably using a forecasting model than a firm that predicts prices, purchasing days, and selling days without a statistical model. A forecasting model is the most accurate prediction method available and is inexpensive to operate once it has been set up.

To demonstrate the value of a price forecasting model for our company, this report (1) documents the effect of price changes on profitability; (2) explains how costs of price forecasting can be minimized; (3) describes two methods of forecasting—moving average models and exponential smoothing models; (4) creates a working model for our company with sample applications; and (5) clarifies the limitations of forecasting. The report demonstrates that the most accurate model for U.S. Packing, Inc. would be one that uses exponential smoothing during periods when market activity is high and switches to a moving average when the market temporarily levels off.

CONTENTS

Figure 13.4 Table of contents.

IMPLICATIONS OF FORECASTING MODELS

Synthesis of a Forecasting System

Clearly, neither model is perfect in the short run. Since the market is generally volatile, exponential smoothing has an inferential advantage. But the market also occasionally experiences periods of stagnation where the moving average performs best.

Ideally, the firm should use exponential smoothing during periods when market activity is high and should switch to a moving average when the market temporarily levels out.

Determination of the Market Activity

Market volatility can be quantified in two ways. The simplest calculation is to summarize the number and amount of all daily fluctuations on each day of the week. Table 4 contains these totals. The second method employs regression analysis, which determines the direction and degree of future market activity for each day of the week.

Forecasting Implications for Mondays

Mondays are traditionally heavy trading days. The mean-receipts column of Table 5 confirms this fact. Despite the negative slope of the best-fitting line through the sample points, this phenomenon appears constant because the correlation coefficient is 0.034.

The large number of livestock transactions seems to have little effect on the live market price. The livestock columns of Table 5 suggest that prices are moderately high and are fairly constant. In fact, Monday's prices have the highest correlation coefficient for the week's livestock prices. This means that a moving average is the best model for price forecasting on Mondays. The livestock columns of

Figure 13.5 Section of forecasting report.

Table 4 Daily Fluctuations

TYPE	Boneless Beef					Livestock					Receipts				
	Mon	Tue	Wed	Thu	Fri	Mon	Tue	Wed	Thu	Fri	Mon	Tue	Wed	Thu	Fri
Number of increases	3	1	3	4	2	2	2	3	2	3	7	3	3	0	8
Number of Decreases	0	0	5	0	0	0	0	3	2	0	0	4	6	9	1
Daily Total	3	1	8	4	2	2	2	6	4	3	7	7	9	9	9
Amount of Increase	3.00	1.50	6.75	2.50	2.50	0.75	1.25	1.50	0.50	1.25					
Amount of Decrease	0.00	0.00	7.00	0.00	0.00	0.00	0.00	1.50	1.25	0.00					
Net change	3.00	1.50	−.25	2.50	2.50	0.75	1.25	0.00	−.75	1.25					

Table 5 Regression Parameters

Type	Boneless Beef					Livestock					Receipts				
	m	b	r	$\overline{X}$	s	m	b	r	$\overline{X}$	s	m	b	r	$\overline{X}$	s
Aggregate	0.30	79.21	0.82	86.47	4.56	−0	26.65	0.04	26.73	1.69	−0	3339	0.03	3206	692
Daily:															
Mon	1.73	78.81	0.90	86.59	4.46	0.65	23.79	0.92	26.72	1.67	−271	6614	0.34	5528	999
Tue	1.62	77.83	0.91	85.96	4.63	0.68	23.35	0.89	26.72	1.96	325	6047	0.32	4422	1585
Wed	1.48	78.54	0.76	85.92	4.62	0.65	23.64	0.89	26.72	1.96	20	3644	0.00	3744	1047
Thu	1.41	79.13	0.74	86.19	4.49	0.55	24.00	0.85	26.75	1.63	−5	658	0.85	638	192
Fri	1.33	78.50	0.82	85.83	4.47	0.57	24.03	0.84	26.89	1.71	−99	2195	0.25	1700	541

Legend

m = slope of the best fitting line through a scatter plot of the period.
b = the Y-intercept of the best-fitting line; indicates price on day zero.
r = correlation coefficient; 1 = perfect fit, 0 = no best-fitting line.
$\overline{X}$ = arithmatic mean for the period.
s = standard deviation for the sample.

Table 4 verify this assertion. Livestock prices fluctuate infrequently, and when they do change, the amplitude of the modulation is small.

Despite the relatively inactive livestock market, the boneless-beef market is moderately active. This suggests that an exponential smoothing model should be used to forecast Monday's boneless-beef prices. Since the correlation coefficient from Table 5 is 0.90, the model should maintain a moderately high tracking speed.

In summary, it is more lucrative to sell boneless beef on Mondays and to utilize livestock inventories rather than buying cattle.

Forecasting Implications for Tuesdays

Tuesday's conditions are a reversal of Monday's. Average boneless-beef prices from Table 5 are considerably lower on Tuesdays. Livestock prices are also reduced. Given these conditions, wholesalers should try to buy livestock on Tuesdays and hold all boneless beef in inventories rather than selling the meat at deflated prices.

Table 4 shows that prices are generally constant on Tuesdays for livestock and boneless beef. Forecasts for both raw inputs prices and boneless beef prices should be made with a moving average model.

Forecasting Implications for Wednesdays

After a brief respite on Tuesdays, institutional buyers return to the market on Wednesdays. As a result, market volatility is expected to increase. Tables 4 and 5 verify this proposition. In addition, Wednesday is the only day of the week that witnesses a net price decrease. This extremely active market does not permit accurate forecasts to be made from either model. However, the exponential smoothing model has a greater probability of success than a moving average if its smoothing constant is large.

Because of the tendency for prices to sharply vacillate on Wednesdays, the decision to buy or sell cannot be made by mathematical analysis. Experience and intuition are the best guidelines for transactions on Wednesdays.

Forecasting Implications for Thursdays

After Wednesday's active trading, livestock surpluses are depleted. Consequently, total receipts drop by more than 3,000 head on Thursdays. The reason is the temporarily high cost per animal. Cattle buyers should avoid large purchases on Thursdays.

Policy implications are reversed for boneless beef. Prices are abnormally high on Thursdays for the finished product. Although the correlation coefficient is only 0.74, wholesalers should attempt to make their largest sales on Thursdays. The relatively low correlation coefficient implies that there is a minor risk associated with the transaction. However, in the long run, the expected yield is large enough to compensate for taking this minor risk.

Table 4 shows that both raw inputs prices and boneless-beef prices are moderately active. This suggests that exponential smoothing should be used to forecast all prices on Thursdays.

Figure 13.5 (*Continued*)

Forecasting Implications for Fridays

Friday's trading is generally moderate. Cattle buyers purchase only enough live-stock to fill Monday's orders. As a result, livestock prices tend to be quite high. If storage facilities are adequate to carry Thursday's stocks over the weekend, Friday's purchases should be kept to a minimum.

Price patterns are normally constant on Fridays. Consequently, moving averages should be used to predict Friday's prices.

All the optimum buying and selling days are summarized in Table 6. The most reliable price forecasting for each individual day also can be found in Table 6.

Table 6   Forecasting Models and Transactions

Day	Boneless Beef		Livestock	
	MODEL	TRANS	MODEL	TRANS
Monday	EXP	SELL	MA	HOLD
Tuesday	MA	HOLD	MA	BUY
Wednesday	EXP	—	EXP	BUY
Thursday	EXP	SELL	EXP	HOLD
Friday	MA	HOLD	MA	HOLD

Legend
MODEL = type of forecasting model.
TRANS = type of transaction.
  EXP = exponential forecasting model.
   MA = moving average forecasting model.
 SELL = conditions are optimum for boneless-beef sales.
  BUY = conditions are optimum for purchasing livestock.
 Hold = conditions are unfavorable—utilize an inventory.

Figure 13.5   (*Continued*)

Glossary

The writer has also included an alphabetical definition of terms so that, for example, a reader unsure of "correlation coefficient" could look up the term and find:

Correlation coefficient: A statistical measurement which indicates how well sample points form a line. If there is a strong linear relationship between the sample points, the coefficient will have a value approaching 1.0. Relative coefficients also have significance. If one group of data has a greater correlation coefficient than another, the points will approximate a line with a greater degree of accuracy.

This report is clear, logical to the reader, and augmented by easy-to-use illustrations and definitions. It anticipates questions and objections, is unlikely to receive negative feedback, and is very likely to be persuasive.

Example 3: An Environmental Impact Statement

Increasingly aware of our limited natural resources, the government has created an Environmental Protection Agency (EPA) that requires all companies contemplating operations that will affect the quality of the environment (air, water, land, and their natural inhabitants) to file an environmental impact statement (EIS). This long report (usually over 1,000 pages) must describe the project's environmental impact, outline a plan to minimize or counteract its negative impact, or, if permanent damage will occur, demonstrate that the need for the contemplated operation outweighs the adverse side effects, or that alternative choices are not feasible or would result in even greater damage. Because of its complexity, preparation of this document requires many problem-solving operations on the part of the writer.

First, the EPA outlines in detail (practically page by page) what must appear and where it must appear. This may seem to simplify the writer's job, in that the outline and format are imposed by the reader. In fact, fitting information into someone else's outline may be difficult to do without distortion, repetition, and/or omission.

Second, to meet the specifications of the outline, reports written by several different writers may be drawn upon by several additional writers, each of whom is responsible for one section of the final EIS. This situation often results in multiple and, sometimes inconsistent use of the same data in different sections. The problem is minimized by stringing annotated page by page outlines of each section along large stretches of board or wall space, where the writers can inspect each others' outlines and correct conflicts in interpretation, or substance before the final report is written.

Third, the draft EIS must be submitted to affected parties in the private and public sectors. Their written comments and any responses the authors of the EIS wish to make are included at the end of the final report. In other words, response to reader feedback becomes formalized. Written responses must answer legitimate questions or complaints but must also be consistent with information in the report itself.

Finally, all the decisions that must be made regarding graphics, transitions, and style are magnified, because the volume of material necessitates making these decisions again and again. Nevertheless, the huge task is cut down to manageable smaller tasks if the writer follows the procedures already explained in this book. Our third example, a two-volume final EIS on the Alternative Fuels Demonstration Program exhibits a strategy very similar to the report-writing technique analyzed earlier in notably shorter examples.

Summary and Format

The two page summary (Figure 13.6) covers in brief all of the major points detailed in over 1,000 pages of text. Frequently, a subject is covered in summary and then gone over again in the same order but in increased detail in a subsequent section. For example, in Volume I, a discussion of "alternatives" to the

SUMMARY

A. Introduction

1. Background

While processes for converting coal, oil shale and other domestic resources into synthetic fuels have been known for many years, they have not been applied on a commercial scale in the U.S. because, in the past, the synthetic products were not competitive in cost with petroleum and natural gas. In order to remedy this situation, a program to accelerate alternative fuels production has been proposed. The three major objectives of this program are:

• to ensure early development of the technical, environmental and economic information and the industry infrastructure needed prior to any possible major expansion of synthetic fuels production capacity
• to improve the nation's international position in energy matters
• to increase domestic energy production in the mid-term by supplementing existing and planned domestic energy production.

2. Program Level Options

This Environmental Statement examines impacts of the program on a plant-by-plant basis and in the aggregate. To examine aggregate impacts, four options for total program implementation have been analyzed. These are as follows:

• *Single phase information option of 350,000 barrels per day by 1985*
  Under this option, synthetic fuels production would be approximately 350,000 barrels a day by 1985. This level was selected with a view toward gaining information about synthetic fuels production and would be attained by constructing one plant of each of the major first generation synthetic fuel types. The information gained from this option would provide significant commercialization experience that could be extrapolated with confidence to a larger commercialization effort if so desired at some future time.

• *Single phase nominal option of 1 million barrels per day by 1985*
  Under this option, synthetic fuels production would be approximately 1 million barrels a day by 1985. This level was selected to balance information gain with the production of a significant amount of usable energy and would be obtained by construction of more than one plant of each type.

• *Two phase nominal option of 1 million barrels per day by 1985*
  Under this option, synthetic fuel production would proceed in two phases. Phase I would be an accelerated development of the Information Option to produce 350,000 barrels a day by 1982. Information developed would be used to decide whether to proceed to Phase II and, if so, to influence the mix of technologies and production schedules in Phase II. Phase II could begin as early as 1978, but would need to be accelerated by 1982 to meet the 1985 production of 1 million barrels a day. This is sometimes called the "shoot-look-shoot" strategy.

Figure 13.6 Environmental impact statement: summary.

- *Single phase maximum production option of 1.7 million barrels per day by 1985*
  Under this option, synthetic fuel production would be 1.7 million barrels a day
  by 1985. This option represents the maximum credible amount of synthetic
  fuels production that could be anticipated with an intense national effort in
  the absence of major dislocations in the economy. It would maximize production of those fuels in shortest supply (e.g., high-Btu gas and petroleum
  substitutes).

This Environmental Statement deals with the potential environmental impacts
that could occur as the result of implementing the Alternative Fuels Demonstration program* under each of these strategies. Three other options are considered
in the Alternative section (see Chapter XII).

3. Synthetic Fuel Technologies

The program to demonstrate the commercial viability of synthetic fuels would
lead to the construction and operation of synthetic fuel plants utilizing technologies that are either now available or nearing readiness for commercialization.
Since one of the major goals of the Program is to develop information on the environmental, economic, institutional, technical and other potential problems associated with the large-scale operation of various types of synthetic fuels plants, it
is expected that several different types of plants would be built, some based on
the conversion of coal to synthetic fuels, some based on the production of synthetic crude from oil shale and others based on the conversion of organic wastes
into synthetic fuels.

Table S-1 lists selected processes that are considered to be likely candidates for
commercial demonstration in the next decade. These processes are currently in
various stages of development and are considered to be representative of the processes that would be included in the program. Detailed descriptions of each
process and its commercialization potential by 1985 are presented in Chapter II.
Processes that are not listed in Table S-1, however, are not precluded from consideration in the program.

4. Approach

This Environmental Statement uses a "building block" approach to assess the
potential environmental impacts of a synthetic fuels industry. This approach
allows for assessing the impacts related to any combination of production levels,
technological mixes and buildup rates.

First, the potential environmental impacts from unit synthetic fuel plants (single plants of a nominal size) are determined for six selected regions: five coal
regions and one oil shale region. Impacts are assessed for unit plants utilizing
each of the processes listed in Table S-1. The potential impacts of conjunctive
developments associated with the unit plants, such as coal mining, community
expansion, pipelines, reservoirs, etc., are then assessed.

*Formerly called the Synthetic Fuels Commercialization Program.

Figure 13.6 *(Continued)*

In order to aggregate the impacts of the unit plants at any production level, alternative industry compositions are developed to determine both the number and the location of each type of unit plant. The aggregate impacts in each region are then determined by combining the impacts of the unit plants and associated conjunctive developments according to the alternative industry compositions.

Table S-1   Candidate Processes for Commercial Demonstration

Coal-Based Technologies

- High-Btu gasification processes
    Fixed-bed process
    Fluidized-bed process
- Low-Btu gasification process
    Fixed-bed process
- Coal liquefaction process
    Fischer-Tropsch process
    Hydrocarbonization process
    Direct catalytic hydrogenation process
    Synthoil process
    Solvent refined coal process

Oil Shale-Based Technologies

- Surface (above ground) process
- Modified in situ process

Solid Waste-Based Technology

- Pyrolysis process

Fuels from Biomass Technologies

- The gasification of biomass (mainly wood and corn stover) to produce synthesis gas for direct utilization as a fuel (medium-Btu gas) or for conversion to synthetic natural gas and/or methyl fuel could be ready for demonstration in the 1988–1990 time frame.

Figure 13.6   (*Continued*)

proposed action appears as part E of the "Introduction" chapter. In Volume II, "Alternatives to the Proposed Program" is a chapter by itself and reexamines the same information in the same order, but in greater detail. (The treatment of solar-energy in both sections has been reproduced in Appendix D so that you can see the different purposes served by the different degrees of coverage.)

Repetition of the same material in different parts of the same report is justified if done for a purpose. In this example, the main point to be made in the "Introduction" is that solar energy cannot compete with the output of the alternative fuels program. In Chapter XII the same point is made, but this time a

As discussed in Chapter XII, Section C.4.a, alternative industry composition C1 was selected for comparison because it would generate the most controlled air emission of any of the six synthetic fuel alternative industry compositions. Similarly, the Mercer County coal was selected as the Western coal for comparison because, on an equivalent-Btu basis, it would generate more of each controlled air emission except particulates, than any of the other western coals considered in the Environmental Statement.

The analysis shows that under the stated assumptions the activities associated with synthetic fuels development and use, as represented by alternative industry composition C1, would generate slightly less sulfur oxide emissions than the activities associated with the direct utilization of a western coal, as represented by a Mercer County, North Dakota coal. (It should also be noted that Tables XII-15 and XII-16 show that all phases of synthetic fuels development generate less particulates, sulfur oxides and nitrogen oxides than would the direct utilization of the selected western coal by itself, even if the emissions from mining and transportation associated with the direct utilization are not included.)

M. North Dakota State Planning Division, State Water Commission

1. Review of Draft Statement

Comment: Report is very general in nature and only repeats information contained in prior reports insofar as environmental matters are concerned. It is difficult to comment on a series of assumptions. If such a plant is proposed for this State, we would certainly have specific comments.

Response: As discussed in the response to comment A.1, this Statement has been prepared so that environmental concerns could be considered in the Program design process at the earliest possible point. At this time it is not possible to prepare site-specific assessments of the Program since specific plants and specific sites have not yet been determined. Detailed site-specific statements would be prepared at the appropriate time for individual synthetic fuel plants that are part of the Program.

N. American Natural Gas Service Company

1. Summary Sheet

Comment: Summary sheet: The reader need go no further than the summary page (second sheet after the cover) to formulate a very unfavorable, and we feel very distorted, impression of the synthetic fuels program. The wording is very negative and unfortunately reflects a tone found throughout the entire document. Statements such as "synthetic fuels conversion plants . . . would create air and water pollution and solid waste and create noise and aesthetic degradation" provide no indication of the magnitude or relative level of expected impact. Providing information on the magnitude of anticipated effects is essential if such a report is to aid in educated decision making.

Figure 13.7 Environmental impact statement: comments and replies.

Response: The Summary Sheet is not meant to provide a detailed summary of the Statement. Rather it is required by the Council on Environmental Quality Guidelines on Preparation of Environmental Impact Statements (40 CFR 1500) and its form and content is prescribed by those Guidelines. It is intended to provide an extremely brief and concise summary concerning the proposed section. Information on the magnitude of anticipated effects is presented in both the body of the Environmental Statement and in the Summary Chapter to the Statement.

2. Hazardous Substances

Comment: Summary Chapter Section D.1.a.(1): The report states that a number of harmful substances may be emitted from synthetic fuels plants. However, no information on the types or concentrations of harmful substances is provided.

Responses: The text clearly indicates that for the most part it is not currently known which, if any, of the potentially toxic materials that may be produced by synthetic fuels plants would be emitted to the atmosphere, nor the amounts that would be emitted. Chapter IV presents a detailed discussion of the potentially harmful substances that may be produced and emitted by synthetic fuels plants.

3. Reclamation

Comment: Summary Chapter, Section D.1.b.(1): The report describes the impacts which would result if reclamation were partially or completely unsuccessful—not the most probable case, which has shown reclamation to be successful if properly performed.

Response: Successful reclamation, under varying conditions and on a scale comparable to that considered in the Environmental Statement, has yet to be demonstrated. The Environmental Statement describes the potential impacts that may result if reclamation is not completely successful or if it results in a change in species composition and/or diversity. The discussion of reclamation in Chapter V, Section A.3.a, has been expanded to provide a more detailed basis for the discussion in this section.

4. Direct Destruction of Wildlife

Comment: Summary Chapter, Section D.1.6.(2): Direct destruction of small mammels by mining is addressed but no mention of a more important consideration is provided, that being the effect in relation to area-wide population of the species.

Response: The effect of direct destruction of some small mammals on the area-wide population of a species would depend upon site-specific conditions and would be strongly infuenced by the cumulative effects of other impacts such as habitat impairment or destruction. It is not possible to discuss the effect of these combined impacts until it is known where the unit plants would be located in each region (see the response to comment A.1).

Figure 13.7 *(Continued)*

257

5. Socioeconomic Impacts

Comment: Summary Chapter, Section D.1.c.(1), third paragraph: Socioeconomic effects appear to be grossly overestimated. Severely overcrowded conditions are anticipated in the report with as much as a quadrupling of present populations. Our studies and those of others indicate such conclusions are inaccurate and extremely unlikely to occur.

Response: This section has been modified as part of the revision and systemization of the socioeconomic impact assessment for the Final Statement. The assumptions behind the potential increases in population are described in Chapter IV.

6. Hyperurbanization

Comment: Summary Chapter, Section D.1.c.(1), second paragraph: The report assumes "hyperurbanization" and boom-town conditions are inevitable. We feel this is an inaccurate assumption that can and will be avoided by advance planning and cooperation between industry and state and local government.

Response: The Environmental Statement does not assume hyperurbanization and boom-town conditions are inevitable. Rather it indicates that there would be a potential for such conditions to occur. Experience in areas such as Gillette and Rock Springs, Wyoming, has shown that such a potential for hyperurbanization does actually exist. It is agreed that advance planning would be required in some areas to prevent or mitigate such an occurrence. This section has been modified as part of the revision of the socioeconomic impact assessment for the Final Statement.

7. Environmental Information Base

Comment: Chapter I, Section D.1, second paragraph: The first sentence, which states that the environmental "information base is not as complete as desired," should be repeated in every case throughout the report where conclusions are based on inadequate data.

Response: Uncertainties and unknowns that are present in the environmental analysis are discussed, where applicable, throughout the Statement. See the responses to comments K.1, K.18 and N.19 for a more detailed discussion of this issue.

8. Conclusions

Comment: Chapter I, Section D.1, third paragraph: The third sentence states that "the (inadequate) knowledge base does not justify excluding any feedstocks or regions of the country from further consideration in the program." However, conclusions are drawn from this inadequate knowledge base throughout the report and this could conceivably have the effect of excluding regions, feedstocks or processes from further consideration.

Figure 13.7 (*Continued*)

258

Response: The findings discussed in the Environmental Statement are based upon the data and analyses presented in the Statement. Assumptions used in the impact analysis are discussed in the Statement and uncertainties and unknowns are pointed out, where applicable.

Figure 13.7   (*Continued*)

full explanation is given of how solar energy operates or might operate in order to prove that it is not a preferable choice.

Reader Feedback

The comments of the American Gas Service Company and the replies of the report writers (Figure 13.7) exemplify the different kinds of responses to reader feedback discussed in Chapter 7. In some cases, the writers explain limitations preventing more precise information (Sections 1 and 4 for instance); in other cases, they have revised the report in response to the reader's reactions (Section 6); and sometimes, they hold their ground (Section 8).

Although it is unlikely that you will every be responsible for writing an EIS or a government contract bid on your own, you might very likely be one of the contributors to it, or even the manager in charge of coordinating everyone else's efforts and giving final approval to the finished product. Knowing that you have mastered a writing process that makes even the *longest* long report manageable will give you the confidence you need to do a good job.

EXERCISES

1. Select a long report (at least 10 pages) from your department file or library on a subject of interest to you. Write a critique of the report using the following checklist as a guide:

   Summary or Abstract: Does it cover problems, methods, solutions, and recommendations adequately?

   Table of contents: Is it accurate, easy to use, and helpful?

   Introduction: Does it prepare you to read the rest?

   Body: Is there a logical movement from general to specific; are headings clear and useful?

   Graphics: Is the design and location of illustrations useful; are necessary graphics missing or unnecessary ones included?

   Conclusion: Does it confirm the introduction, follow from the body, and leave you with a positive feeling about the report?

2. Following the guidelines in this book and the advice of your instructor, create a long technical report (at least 10 pages), preferably on a project you are engaged in at work or in one of your other classes. Get reader feedback from your fellow students. If they or your instructor have any comments with which you disagree, see if you can formulate a persuasive written response.

CHAPTER 14

Preparing and Delivering Oral Reports

Whatever your profession, you will experience several circumstances requiring you to preface or back up what you write with what you say. The three steps to successful oral communication of technical material are:

- *Know* your subject well.
- *Plan* your presentation structure carefully.
- *Practice* your delivery until you feel a confidence and ease that you can convey to your audience.

To accomplish these steps, you will apply most of the same packaging procedures that you have learned for effective technical writing.

SITUATIONS REQUIRING ORAL REPORTS

One or more of the following professional situations will require that you know how to present technical information orally:

1. A VIP in your own firm or a client firm, a government official, or an insurance investigator is visiting your company's facilities. You are asked to show the visitor around, provide a running commentary on how everything works, and answer the visitor's questions.
2. You must visit a potential client's or customer's office and make a presentation to the Board of Directors, none of whom you have ever met before, that will convince them to sign a contract for your company's product or services.
3. You must present a technical paper at a professional organization's convention.
4. You are asked to participate in a training program for new personnel.

5. You are asked to testify before a government agency on behalf of your organization in order to request permission to pursue research and development operations that might have an adverse affect on the environment, or in order to lobby for decisions that might promote or favor the needs of participants in your industry or residents of the community in which your business is located.

6. You are asked to participate in your company's speakers' bureau and provide informative talks on how what your company does affects the local or national residential or business community.

7. You need to give an oral summary of a proposal, or progress report, or an oral briefing on a technical report at a department, division, section or project meeting.

KNOW YOUR SUBJECT

Some oral communications are written communications that are read out loud. Others are delivered using notes; and still others are given extemporaneously, or "off the top of your head," with seemingly little advance preparation. Whichever of these three approaches you find appropriate to your particular situation, the most important impression you need to convey is that you know your material. Usually, after your first few sentences, your audience has decided whether you know what you are talking about and are someone worthy of their close attention. With few exceptions, this initial impression is rarely reversed, and frequently, the time you have to make that impression is limited.

For example, an industry or institutional representative at a Senate committee hearing might submit a detailed written testimony prior to the hearing. Unfortunately, few if any of the committee participants will have had time to more than glance at it before they hear the oral summary. Presenters have about five minutes to introduce and explain their main points to listeners who are relatively unfamiliar with the details and who, in addition, may have interfering preconceptions about what you will say and why.

In typical situations such as these, you must be well enough acquainted with your subject to:

1. Present it precisely and succinctly.
2. Adapt the format and style of your presentation to the needs and expectations of your listeners.
3. Restate and further substantiate your main points throughout your delivery and in your responses to questions and comments.

The best method of knowing your subject well enough to accomplish your task is the same technique that you use for organizing short and long written communications. Make a list or outline of everything you want to cover in your presentations, and set it out in front of you so that you can work with it.

PLAN YOUR PRESENTATION

When you are using a list or outline to determine an effective format for an oral presentation, you will want to keep in mind that while a reading audience can reread or re-examine a section that they missed the full import of, a listening audience usually gets just one chance to grasp your meaning. Therefore, particularly if the material is complex and unfamiliar, you should plan to repeat or rephrase your salient points at least two or three times—minimally, at the beginning, middle and end of your talk—even if your presentation is relatively short.

To be instructively repetitive without being tedious, you will need to make accurate judgments about the level of competence and the special interests of those to whom you are speaking. Know your audience's point of view towards your material and utilize the appropriate adaptation of your material to their point of view to accomplish the purpose you want your oral communication to serve.

Example: Preparing and Anticipating

The assistant manager of Product Manufacturing in Firm X is responsible for adapting current plant procedures to the use of two new pieces of equipment recently purchased by the head of Product Engineering from a foreign manufacturer. He is scheduled to give an oral report outlining how the equipment will change, and hopefully improve, the manufacture of product line B, and what he is doing to make the changes smooth and efficient. His audience will be his supervisor, the head of Product Manufacturing; his supervisor's boss, the Vice President of the division to which Product Manufacturing belongs; and appropriate representatives from Product Design, Advanced Mechanical Engineering, Sales, and Accounting.

There is one special problem. The assistant manager realizes that in the budget for purchasing his supervisor did not include a sum of money to pay for two engineers to travel to the country where the new equipment is manufactured and to participate in a training program. To the assistant manager, leaving training money out of the budget was a serious mistake. Without special training, the company can make only minimal use of the expensive new equipment. They will be limited to adapting current procedures to the new machines. If they were to invest the small sum necessary to train two engineers in all of the equipment's potential uses in their industry,

- the two could come back and train all other users of the equipment
- the profit from purchase of the machines could be expanded to include development of new methods and applications in addition to adaptation of existing procedures.

The assistant manager has the figures to back up his claim that spending money on training is more cost efficient over the long run than not spending money on training. These figures will probably convince most of the people scheduled to attend the meeting.

But what about his supervisor? If, indeed, training is preferable to no training, then the head of Product Manufacturing should have included cost of training in his original projections of how much the purchase of the new equipment would cost. Therefore, it will be difficult for the assistant manager to include in his oral report a justified request for training money without implying that his supervisor was remiss in not making a similar request in the first place.

Understanding this organizational situation, the assistant manager realizes that the strategy of his oral presentation must incorporate three goals:

1. To convince upper management representatives that spending money on training in use of the new equipment is cost efficient.
2. To anticipate and answer any objections raised by his supervisor, who will naturally wish to defend his position of not including training in the original purchasing budget.
3. To avoid offending or attacking his supervisor while still presenting a plan that changes the original one.

Probably, the speaker will begin by reviewing in a positive way the original reasons for purchasing the equipment (e.g., improving efficiency in the manufacture of Product Line B). He will explain, using charts and diagrams, how these changes will be implemented, and what the payback period will be.

Then, he will suggest that as long as the equipment is in-house, it occurred to him that the company might wish to explore other possible uses for it. He will suggest some specific possibilities and mention that the equipment company does not have a training program for this purpose.

He will describe the training program and recommend enrolling two engineers. He will be prepared to answer questions such as how the two engineers' ongoing projects would be continued while they are gone.

Also, if he is smart, he will make a point of demonstrating that the positive effects of utilizing this training program are in keeping with the spirit of his supervisor's original plan of increasing manufacturing efficiency in his division.

PRACTICE YOUR DELIVERY

Practice makes perfect. But before you can practice anything, you need to have standards or criteria against which to measure your performance. You cannot work toward presenting an effective speech through practice, until you have decided what an effective speech is.

Perhaps the best way to learn what works in communicating orally is to become scrupulously analytical of the methodologies of everyone who speaks to you. Any time anyone presents information to you orally—teachers, PTA officers, clergymen, news commentators—even door-to-door salespersons—evaluate what is taking place in the speaker/listener dynamic:

- If you find yourself listening with interest, ask yourself why.
- If the speaker's main points are easy to identify, understand, relate to one

another, and remember, what techniques of oral presentation are responsible?

- Does the speaker have personal mannerisms that call attention to how he or she is speaking, and away from the points he or she is making? If so, what might the speaker do to prevent or minimize these distractions?

- Does the speaker repeat the same or similar points too often or too little? How much and what kind of repetition is effective as a speaking tool?

- Does the speaker allow technical difficulties (e.g., squeaky microphones; awkwardness at handling slides, overhead projectors, sheets of paper, etc.) to undermine or diminish the effectiveness of the presentation? If so, what might he or she have done to anticipate such inconveniences or to be prepared to deal with them on the spot, if they cannot be anticipated?

- Does the speaker respond confidently and clearly to the audience's comments and questions? If not, what does he or she need to do to improve?

- Given the physical environment, the subject and the level of audience, is the speaker's delivery too long, too short, or just right?

- Has the speaker adapted to the anticipated expectations and needs of the audience as well as possible?

- What responses do you hear from other members of the audience as they depart?

Evaluating the oral presentations of others is an ideal method of learning what makes oral communication effective under various circumstances, and of adapting the successful practices of others to the oral communication requirements of your own position.

Once you know what effects you are aiming for, you are ready to practice. *Practice alone* first. If possible, approximate the physical circumstances you will be in (practice in a large room, or use a podium or a pointer).

If you can, *record* your practice session. Then you can play it back and hear what you sound like to others. If you will be using a microphone, try to practice with one. For example, if you tend to wave your hands around when you talk, learn to avoid hitting the mike when you punctuate a point with a gesture.

If you or your company has access to videotape equipment, you may want to videotape a practice session. Once you have seen yourself sway back and forth, scratch your head or wiggle your fingers every third sentence, you will be cured of these distracting gestures, forever. Finally, practice in front of others, if possible, to get live audience feedback on your performance.

Again, your end goal is to project, in your content and in your delivery, a command of your subject and a confidence in yourself and in what you have to say. People believe in people who believe in themselves. Self-assurance and enthusiasm are contagious.

MASTERING THE ORAL DYNAMICS OF GROUP CONFERENCES

The advantage of being a key speaker whether at a convention, a stockholder's meeting, or a sales presentation is that for an agreed-upon period of time, you have the floor—the audience's guaranteed attention. You know beforehand what the subject will be, and you can prepare accordingly. All the givens are on your side.

In the course of your career, however, you will frequently be in less formal situations where oral communication skills are vital. In business, groups of people often get together to discuss a project. Each member of the group is jockeying for position, trying to get his or her point across, and competing with others for a key party's attention. Put in this situation, you neither want to monopolize the conversation, nor to sit silently in your corner. How do you steer between these extremes? How do you make your point without antagonizing others in the group?

1. Know the purpose of the meeting and what purpose you want it to serve for you.
2. Begin by listening. Let others speak and really hear what they are saying so that you can feel out (a) when the time is right for you to speak up, and (b) what strategy they are capable of responding to.
3. Come in with your main point only when you feel that the group is ready— only after you have let others set the stage for you.

After you have attended a few company meetings and have studied the strategies of the regular participants, you will be able to judge the right time to speak— after the tension has built and before it has died. But meeting dynamics are worth your observation and analysis because being able to control what happens in verbal interchanges is an effective way of calling attention to yourself as an efficient, confident and valuable member of your organization.

EXERCISES

1. Record and analyze the oral dynamics of a class session or a meeting of a group to which you belong. Consider the strategies, timing, and attitudes of each of the people who spoke. What would you do differently? Why? How?
2. Attend a public lecture and analyze the effective and/or ineffective techniques of the speakers.
3. Prepare an oral presentation on a school or community issue and give it to your class or to a group of friends.

Appendix A
Report on Cost Control Procedures in the Restaurant Business

COST CONTROL PROCEDURES

Many different approaches have been taken over the years in an effort to control food costs. However, the objective has always been the same: to keep costs in line with what they should be without sacrificing the quality or quantity of the food which goes to the customer.

Two of the most effective cost control procedures are Standards and Planned Production.

Standards

The word "standards" is synonymous with the phrase "what something should be." Standards are guides to action and aids to those responsible for achieving the organization's desired objectives. Four different standards are discussed:

Standard portion sizes

Standard recipes

Standard purchase specifications

Standard yields.

Each standard is important by itself, but all must be implemented to achieve results.

Standard Portion Sizes

Standard portion sizes represent the number of ounces of each food item which is sold to the club's patrons for a stated price. Standard portion sizes are established for all items—appetizers through desserts.

Standard portion sizes are important for two main reasons. First, the customers of the club should be served the quantity of food for which they are paying. Second, if the menu price is set for nine ounces of strip steak, and if ten ounces are served, the club is the loser.

An example of a standard portion size taken from the club's dinner menu follows:

> *Scampies*
> 4-U-10 Scampies $3.00
> 2 oz. Butter   .04
> Parsley   .01
> Surrounding Plate Cost   .55
> $3.60

Standard Recipes

A standard recipe is a written formula prescribed for producing a food item of a specified and desired quality and quantity. The recipes show the exact quality of each ingredient used in making the item and the sequence of steps to be followed in its preparation.

Standard recipes serve three primary purposes. They aid in ensuring consistency in quality food preparation; in determining standard food costs; and in determining the raw food cost to prepare the recipe item so that it may be properly priced. Figure 1 is an example of a standard recipe.

Standard Yields

Yield means the net weight or volume of a food item after it has been processed and made ready for sale to the guest. Standard yields are those that result when an item is processed according to established standard preparation procedures.

Yields serve two major functions. First, they are important in determining the selling price of an item. Second, they provide management with a potential dollar amount of income per item based upon how many portions (yield) can be carved from that item.

An example of a yield test can be seen in Figure 2.

Calculation of Standard Food Costs

The standard cost of food sold represents an ideal cost that would be realized if (1) there were no waste, (2) there were optimum efficiency, and (3) all supporting standards were followed. Since this cost is an ideal, it is necessary to add in a tolerance factor to allow for unavoidable waste and inefficiency. This adjusted cost becomes the practical food cost standard—literally what the cost should be.

The primary purpose of a standard cost is to serve as a comparison basis for evaluating actual food cost results. It gives management the information necessary to measure the operational efficiency of its staff with respect to food costs as the month goes by. Without such information, management would have no way of knowing whether the actual cost of food was in line with what the cost should be. If what the cost should be is not known, the door is left open for an indeterminate amount of loss from inefficiency, avoidable waste and possible theft.

INGREDIENTS	QUANTITY	METHOD FOR PREPARING
Boneless Rib Eye of Beef (small eye) Or Strip Loin	12 lb.	1. Cut 25 6–7 oz. steaks from rib eye. Refrigerate. 2. Immediately prior to service, remove steaks from refrigerator and saute in a hot pan or cook on a hot grill. Cook to desired doneness and place on warm individual plates or platters. 3. Place a slice of Bercy Butter (1 oz.) on top of hot steak, serve immediately. Garnish with a sprig of watercress or fresh parsley.
Bercy Butter:		
Butter	1½ lb.	1. Cream butter until smooth. Add
Shallots, minced fine	1 tbsp.	shallots, chives, parsley and
Chives, minced fine	1 tsp.	tarragon. Form a long roll about as
Parsley, minced fine	1 tbsp.	big around as a 50-cent piece. Wrap
Tarragon Leaves, minced fine	1 tsp.	firmly in waxed paper and refrigerate until service time.

Figure 1. Standard Recipe for Entrecote Bercy (25 portions, each 6–7 oz. with 1 oz. Bercy Butter)

To calculate the standard food cost for the entire Club operation, three types of facts are required:

1. An accurate count of the individual menu items sold.
2. The complete cost, including surrounding dish costs, for all food items used.
3. The sales prices of all items on the menu.

Figure 2. Summary of Yield Test Results

Item: Beef Rib, Oven Prepared	Grade: Choice
Item Cost: $28.28 at $1.40/lb.	Weight: 20 lb., 4 oz.

Summary of Yield Test

Cooking and Carving Details	Weight	Percentage of Original Weight	Cost per Servable Pound
Servable Weight	11 lb. 3 oz.	55.2	$2.53
Loss in Carving	5 lb. 3 oz.	25.6	
Loss in Cooking	3 lb. 14 oz.	19.2	
	20 lb. 4 oz.	100.0	$1.40

A sample worksheet of standard food costs for the Club's seafood entrees can be found in Figure 3.

41.5 represents an ideal cost percentage. As mentioned earlier, it is necessary to add an allowance or tolerance to the ideal standard cost percentage in order to arrive at a practical standard. In the case of the Club, a 1.5 percent tolerance factor is used, which results in a practical standard cost of food sold of 43 percent.

Production Planning

Proper food production planning is one of the most effective methods for avoiding food waste. Production planning should be designed to determine food production requirements in advance of preparation, based on known and/or anticipated sales. The information required to attain the objective of production is of two kinds:

1. An accurate estimate of the number of portions of each item which will be consumed.
2. The number of raw pounds of merchandise required to produce the number of forecasted portions.

To accurately forecast the number of portions to be consumed, it is necessary to have some record of what was consumed in the past. One of the most effective ways of gathering this data is by means of a Sales History Record. Based on the belief that the future will be a reflection of the past, a percentage of total sales (for each item) during a given time period can be determined. The formula is

$$\frac{\text{Number of Item Sold}}{\text{Total Customers Served}} = \text{Percentage of Total Sales}$$

If the total number of customers for a given day is accurately estimated, the expected number of each menu item may be determined by multiplying the recorded percentages by the total number of customers forecasted. The results of this calculation provide a reliable guide for estimating the number of portions to be served.

Figure 3. Worksheet for Calculation of Standard Cost of Food Sold during June, 1977

Menu Item	Total Item Cost	Sales Price	Number Sold	Weighted Cost Value	Weighted Sales Value	Standard Food Cost Percentage
Entrees						
Snow Crab Claws	$2.30	$5.75	12	$ 27.60	$ 69.00	40.0
Scampies	3.60	6.95	48	172.80	333.60	51.7
Bay Scallops	1.65	4.75	32	52.80	152.00	34.7
King Crab Legs	2.86	6.45	23	65.78	148.35	44.5
Scrod	1.31	4.25	73	95.63	310.25	30.8
Dover Sole	3.90	7.25	6	23.40	43.50	53.7
Total				$438.01	$1056.70	41.5

The raw material required (to sustain the estimated demand) will depend on the size of the portion to be served, the purchase specification of the item, and the yield of the item after preparation.

From this information, a food production planning worksheet can be prepared. This worksheet is used for three primary reasons: First, it provides for an easy summarization for the forecast data. Second, it provides a ready data bank for the chef to determine his or her raw material requirements. Third, it provides space to record the actual results of the meal period, which contributes to easier future analysis.

Appendix B
NASA Technical Note

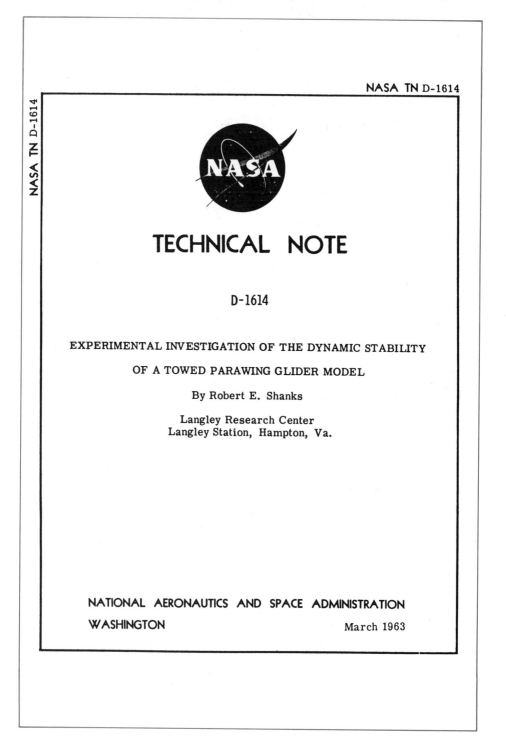

NASA TN D-1614

NASA TN D-1614

TECHNICAL NOTE

D-1614

EXPERIMENTAL INVESTIGATION OF THE DYNAMIC STABILITY
OF A TOWED PARAWING GLIDER MODEL

By Robert E. Shanks

Langley Research Center
Langley Station, Hampton, Va.

NATIONAL AERONAUTICS AND SPACE ADMINISTRATION
WASHINGTON   March 1963

NATIONAL AERONAUTICS AND SPACE ADMINISTRATION

TECHNICAL NOTE D-1614

EXPERIMENTAL INVESTIGATION OF THE DYNAMIC STABILITY

OF A TOWED PARAWING GLIDER MODEL

By Robert E. Shanks

SUMMARY

An investigation of the dynamic stability characteristics of a towed para-wing glider has been made in the Langley full-scale tunnel by means of free-flight model tests. The model was tested in its basic configuration (wing, risers, and a steel weight representing a payload) and with various amounts of vertical side area added in the plane of symmetry beneath the wing.

The investigation showed that the basic configuration had unsatisfactory tow characteristics because of a large, constant-amplitude lateral oscillation which seemed to consist of a large amount of sidewise motion in proportion to the roll and yaw. It was found that the addition of vertical side area could provide satisfactory tow characteristics if the area was properly located. It should be pointed out that in this investigation the model did not have a cargo package of any appreciable dimensions and, therefore, only the aerodynamics of the wing and vertical panels were involved. For other configurations a somewhat different arrangement of side area might be necessary to achieve satisfactory tow.

INTRODUCTION

The use of towed gliders for transporting troops and material has been the subject of much study for a number of years. In practice it has generally been found necessary for the towed vehicle to be piloted because of difficulty in achieving an inherently stable tow configuration. Inasmuch as use of the para-wing as an unmanned cargo-carrying towed glider is being studied by the military, an investigation is being conducted by the National Aeronautics and Space Administration in an effort to determine a satisfactory tow configuration.

The tow tests of the investigation were conducted in the Langley full-scale tunnel to determine the dynamic stability characteristics of a towed parawing model. A simple model in which the cargo was simulated by a steel weight suspended by rigid members below the wing was used in this study. The effects of various changes in geometry, which were achieved by adding vertical side area in the plane of symmetry below the keel, were investigated during the program. Static force tests were made in a low-speed tunnel with a 12-foot octagonal test section at the Langley Research Center over an angle-of-attack range from 20° to 40° in order to determine the static stability characteristics of the various configurations.

SYMBOLS

All lateral data are referred to the body system of axes (fig. 1) and the longitudinal data are referred to the wind axes. All coefficients are based on the flat pattern area of 17.7 square feet, a keel length of 5.0 feet, and a span of 7.1 feet. The moments are referred to the reference center-of-gravity position.

b	wing span (flat pattern), ft
l_k	keel length, ft
F_D	drag force, lb
F_Y	side force, lb
I_X, I_Y, I_Z	moment of inertia about X-, Y-, and Z-axis, respectively, slug-ft^2
F_L	lift force, lb
L/D	lift-drag ratio
M_X	rolling moment, ft-lb
M_Y	pitching moment, ft-lb
M_Z	yawing moment, ft-lb
q	free-stream dynamic pressure, lb/sq ft
S	wing area, sq ft
X, Y, Z	coordinates axes
α	angle of attack of keel, deg
β	angle of sideslip, deg or radians
C_D	drag coefficient, F_D/qS

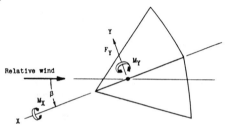

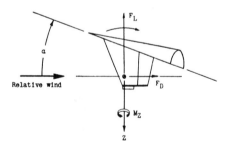

Figure 1.- Sketch of axis systems used in investigation. Arrows indicate positive directions of forces, moments, and angles.

C_L lift coefficient F_L/qS

C_l rolling-moment coefficient, M_X/qSb

C_m pitching-moment coefficient, M_Y/qSl_k

C_n yawing-moment coefficient, M_Z/qSb

C_Y side-force coefficient, F_Y/qS

$C_{l_\beta} = \dfrac{\partial C_l}{\partial \beta}$ per deg

$C_{n_\beta} = \dfrac{\partial C_n}{\partial \beta}$ per deg

$C_{Y_\beta} = \dfrac{\partial C_Y}{\partial \beta}$ per deg

APPARATUS AND TESTING TECHNIQUE

Model

The model used in the investigation was constructed at the Langley Research Center. The wing leading edges and keel consisted of 5-foot lengths of aluminum tubing, 0.75 inch in diameter. The leading edges were fixed at 50° sweep by means of a spreader bar. The fabric was nonporous Mylar bonded to ripstop nylon and had a canopy flat-pattern sweep of 45°. Rigid risers of 0.5-inch-diameter aluminum tubing attached to the leading edges and keel supported an 8-pound steel bar which represented the weight of a payload. Details of the model are given in figure 2 and table I. In its basic configuration the model consisted only of the wing, risers, and steel weight as shown in figure 2(a). It was also provided, however, with the various vertical

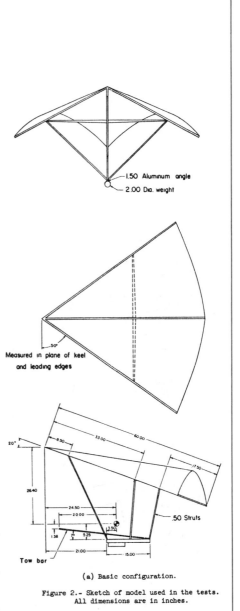

(a) Basic configuration.

Figure 2.- Sketch of model used in the tests. All dimensions are in inches.

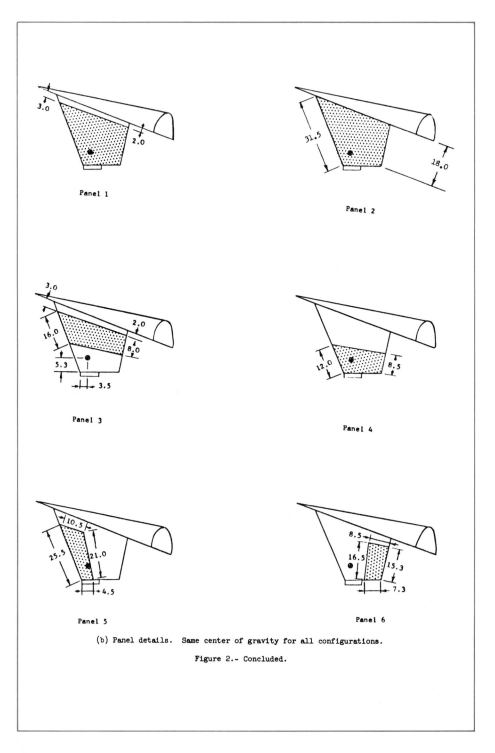

Panel 1

Panel 2

Panel 3

Panel 4

Panel 5

Panel 6

(b) Panel details. Same center of gravity for all configurations.

Figure 2.- Concluded.

TABLE I.- MASS AND GEOMETRIC CHARACTERISTICS OF THE MODEL

Weight, lb 13.9

Moments of inertia:
I_X, slug-ft^2 . . . .   0.77
I_Y, slug-ft^2 . . . .   0.55
I_Z, slug-ft^2 . . . .   0.36

Vertical-panel areas, sq ft:
Panel 1 . . . .   3.1
Panel 2 . . . .   3.6
Panel 3 . . . .   1.9
Panel 4 . . . .   1.2
Panel 5 . . . .   1.0
Panel 6 . . . .   0.9

Wing:	Flat pattern	Flight
Sweep, deg	45	50
Area, sq ft	17.7	16.0
Span, ft	7.1	6.4
Aspect ratio	2.8	2.6
Root chord (keel length), ft	5.0	5.0

panels shown in figure 2(b), which were located beneath the keel in the plane of symmetry.

Test Equipment and Setup

The force tests were conducted in a low-speed tunnel having a 12-foot octagonal test section at the Langley Research Center. The model was sting mounted, and the forces and moments were measured about the body axes by using strain-gage balances.

Flight tests to study the dynamic stability of the model when towed were conducted in the Langley full-scale tunnel with the test setup illustrated in figure 3. A 1/32-inch-diameter aircraft cable towline was attached to the turning vanes ahead of the tunnel contraction. This arrangement resulted in a towline length of 140 feet. An overhead safety cable was used to restrain the model from excessively large motions and was handled by an operator who kept it slack during flight or took up the slack to prevent the model from crashing if its motions became too violent. Motion-picture records were obtained with a camera located at the three-quarter rear location in the test section.

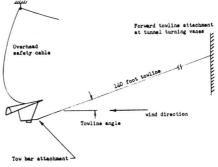

Figure 3.- Test setup used for towing the model.

In this investigation the flights were started with the model hanging on the safety cable. The tunnel speed was then brought up to that required for the particular trim conditions and the model would lift off the safety cable. After the flight behavior had been studied for the required period of time the flight was terminated by decreasing the tunnel speed and taking up the slack in the safety cable.

FLIGHT TESTS

Flight tests were made to determine the dynamic stability characteristics of the model in towed flight. The model was tested in its basic condition and with various vertical panels located in the plane of symmetry beneath the keel. The flights were made at a keel angle of attack of about $20°$ which corresponds approximately to the angle of attack for maximum lift-drag ratio. The tunnel speed for the tow tests was 24 miles per hour. The towline was attached to the rigid towbar shown in figure 2 so that the towline tension acted approximately through the center of gravity. The effect of towline attachment point was investigated but most of the tests were run with the towline attached at the end of the towbar shown in figure 2.

STABILITY PARAMETERS OF THE MODEL

In order to aid in the interpretation of the flight-test results, force tests were made to determine the static longitudinal and lateral stability characteristics of the model that was flight tested. The tests were run at a dynamic pressure of 1.6 pounds per square foot which corresponds to an airspeed of 37 feet per second at standard sea-level conditions and a test Reynolds number of 1.18×10^6 based on the keel length of 5 feet.

Static Longitudinal Stability

The static longitudinal stability tests were made for an angle-of-attack range from $15°$ to $40°$ for the basic model and for the model with panel 1 on. These data are presented in figure 4 and show virtually no effect of the panel on the longitudinal characteristics, as might be expected.

Static Lateral Stability

The static lateral stability characteristics of the model were determined over a keel angle-of-attack range from $20°$ to $40°$ for a sideslip range up to $\pm20°$ The results of these tests are presented in figure 5. These data are summarized in figures 6 and 7 and in the form of the stability derivatives C_{Y_β}, C_{n_β}, and C_{l_β} plotted against angle of attack. The values of these derivatives were obtained from the difference between the values of the coefficients measured at

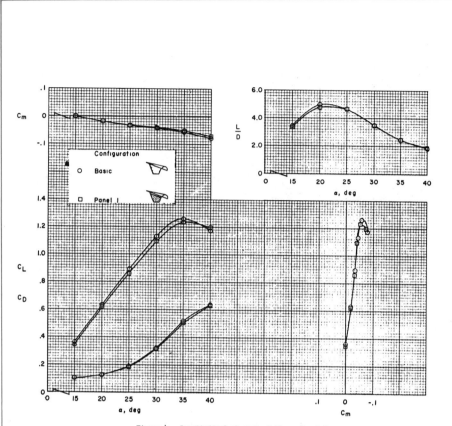

Figure 4.- Longitudinal characteristics of model.

sideslip angles of 5° and -5°. Since the data of figure 5 show nonlinearity at some angles of attack, the derivatives presented in figures 6 and 7 are only used to indicate trends and to provide approximate comparisons of various configurations.

In general, the data of figure 6 show that panels 1 and 2 had some rather large effects on the values of the lateral derivatives. Examination of the panel details presented in figure 2(b) shows that the changes in the derivatives can be accounted for by the changes in the distribution of the panel area with respect to the center of gravity and by the presence of the slot between panel 1 and the wing to reduce the end-plate effect of the wing on the panel. The stability derivatives for the model with panels 1, 3, 4, 5, and 6 on are compared in figure 7 and the effects of the panels are generally those that would be expected.

FLIGHT-TEST RESULTS AND DISCUSSION

A motion-picture film supplement covering flight tests of the model has been prepared and is available on loan. A request card form and a description of the film will be found at the back of this paper on the page immediately preceding the abstract and index page.

If the simplified test model was assumed to be 0.25 scale, the model represented a full-scale glider having a keel length of 20 feet, a wing loading of 3.1 pounds per square foot, a tow speed of 48 miles per hour, and a towline length of 560 feet. The scaled-up values presented were obtained by using the dynamic similarity relationships which are summarized in some detail in reference 1.

Basic Configuration

The tow tests showed that the behavior of the model in the basic configuration was unsatisfactory because of a lateral oscillation which usually appeared to be a constant-amplitude (5 or 6 span lengths) lateral translation back and forth across the test section with relatively small amplitudes in roll and yaw. The oscillation was very sensitive to gusts and other disturbances and at various times would appear to be stable, neutrally stable, or unstable depending on the disturbance striking the model. Occasionally, the oscillation would be abruptly damped by a disturbance and the model would appear to be in stable tow for a short time. However, another disturbance would soon trigger the oscillation so that it would build up again and continue for long periods of time. The fact that the oscillation generally appeared to be of relatively large and constant amplitude was taken to indicate that the oscillation was unstable for small amplitudes; and the fact that it was sometimes damped by disturbances and did not build up immediately was taken to indicate that the degree of instability was low.

The longitudinal characteristics were generally satisfactory. There was some vertical movement of the model but the motions were slow and random in nature and of fairly small amplitude (1 or 2 span lengths).

A number of exploratory tests were made in which vertical and horizontal attachment points and towline angle were varied in an effort to improve the towing characteristics. It was found that the best stability was achieved when the towline action was approximately through the center of gravity. Also, the greater the towline angle for the range tested (10° to 15°), with the model lower than the attachment point on the tunnel turning vanes, the easier it was to achieve a stable condition. A large towline angle, however, requires the towing aircraft to operate at a higher power since it must provide a large portion of the lift force for the glider. In this investigation most flights were made with a towline angle of about 12°. Although increasing the towbar length about 50 percent resulted in slightly better stability characteristics, the improvement was not great enough to warrant an extremely long towbar; thus, all subsequent tests were made with the 20-inch (1/3 keel length) towbar shown in figure 2(a).

With the towline attachment point and the towline angle established in this part of the investigation, additional tests were made in which various means were tried in an effort to improve further the lateral stability of the model. From these studies it was found that adding vertical side area in the plane of symmetry beneath the wing could provide a pronounced improvement in the damping of the oscillation, but the location of the area was important. The remainder of the discussion will deal with the effect of area size and location on the lateral stability.

Revised Configurations

Panel 1.- The best lateral stability characteristics were achieved with panel 1. (See fig. 2.) With this configuration the model was generally very steady and there was virtually no translational motion. Occasionally, the model was upset by gusts and disturbances but the damping of the ensuing oscillation was almost deadbeat. The overall characteristics of this configuration were considered to be excellent.

Panel 2.- The lateral stability of the model was greatly changed by the addition of the small strip of area directly beneath the keel that was the only difference between panels 1 and 2. (See fig. 2.) The translational motion was virtually eliminated but a short-period (about 1-second) constant-amplitude, Dutch roll type of oscillation about the towbar attachment point appeared. There was sufficient energy in the oscillation that it was relatively unaffected by gusts and other disturbances; thus there was never any tendency to change the character of the motion. The characteristics of this configuration were considered to be unsatisfactory.

The appearance of the Dutch roll oscillation for the configuration with panel 2 and not for that with panel 1 can probably be explained by the decrease in directional stability $\left(C_{n_\beta}\right)$ and the increase in effective dihedral $\left(-C_{l_\beta}\right)$ (see figs. 6 and 7), derivative changes that greatly reduce the damping of the Dutch roll oscillation of a free-flying airplane. Since it has been pointed out in reference 2 that the addition of a towline to an airplane usually had little effect on the Dutch roll mode, it appears that the Dutch roll oscillation experienced for the panel 2 configuration was due to the change in C_{n_β} and $-C_{l_\beta}$.

Panels 3 and 4.- When the side area of the vertical panel was reduced considerably from that of panel 1 to that of panels 3 and 4, the lateral motions of the model were generally similar to those of the basic configuration although the motion was lightly damped and there were longer periods of steady flight than with the basic configuration.

Panels 5 and 6.- A reduction in the side area in the fore-and-aft direction did not result in satisfactory tow characteristics but the behavior of the model varied greatly with location of the side area, evidently because of differences

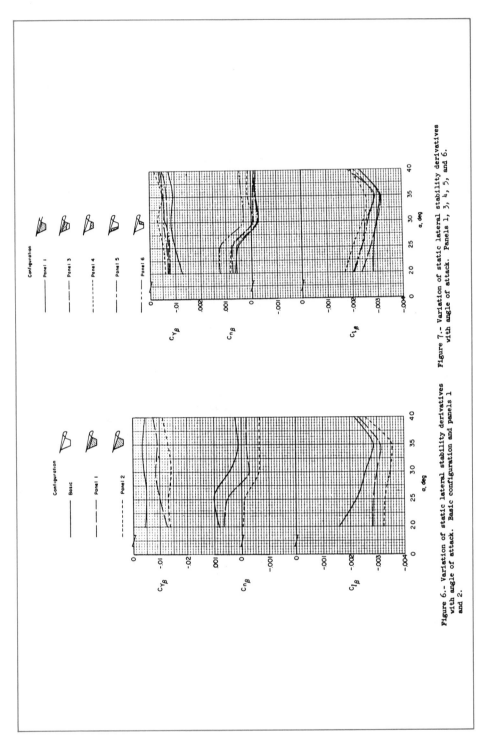

Figure 6.- Variation of static lateral stability derivatives with angle of attack. Basic configuration and panels 1 and 2.

Figure 7.- Variation of static lateral stability derivatives with angle of attack. Panels 1, 3, 4, 5, and 6.

in directional stability between panels 5 and 6. With panel 5 the tow characteristics were actually fairly good. The model had a tendency toward a lateral
oscillation similar to that of the basic configuration but there was definitely
a damping tendency at times and long periods of steady flight. With panel 6,
however, the model had the large constant-amplitude oscillation characteristic
of the basic configuration. In fact, the oscillation appeared to be worse
because it never damped out. Also, the model had a strong weathercock tendency,
because of the increased directional stability, which resulted in considerable
yawing displacement of the model about the towbar attachment point during
translation.

CONCLUDING REMARKS

An investigation in the Langley full-scale tunnel to study the tow characteristics of a parawing glider indicated that the basic configuration (wing,
risers, and a steel weight representing a payload) was unsatisfactory because
of a constant-amplitude lateral oscillation which appeared mainly as sidewise
motion. It was found that the addition of vertical side area in the plane of
symmetry beneath the wing made the tow characteristics satisfactory, if the area
was properly located. It should be pointed out that in this investigation the
model did not have a cargo package of any appreciable dimensions and that, therefore, only the aerodynamics of the wing and vertical panels were involved. For
other configurations and angles of attack, a somewhat different arrangement of
side area might be necessary to achieve satisfactory tow.

Langley Research Center,
     National Aeronautics and Space Administration,
          Langley Station, Hampton, Va., November 28, 1962.

REFERENCES

1. Shortal, Joseph A., and Osterhout, Clayton J.: Preliminary Stability and
     Control Tests in the NACA Free-Flight Wind Tunnel and Correlation With
     Full-Scale Flight Tests. NACA TN 810, 1941.

2. Schy, Albert A., and Woodling, Carroll H.: Preliminary Theoretical Investiga-
     tion of Several Methods for Stabilizing the Lateral Motion of a High-Speed
     Fighter Airplane Towed by a Single Cable. NACA RM L52L24, 1953.

Appendix C
INTRODUCTION TO THE DECSYSTEM-20 A. R. Jennings Computing Center

Table of Contents

This document introduces you to ARJCC's DECsystem-20 (running under TOPS-20). It explains simple commands for logging in and out, editing, and running programs. It is assumed that you are a novice, but that you understand the general functions of computers. All serious users should acquire the DEC User's Guide (available at the Book Store) to learn more about the topics presented here. Specialized manuals for different programming languages, the editor, and other topics are also available.

You need a user name, a password, and an account name to use the system. Your instructor or department should help you get these. (Keep your password confidential so that unauthorized people cannot use your computer time.)

The first time you use the system, you should use a terminal that is connected directly to the computer rather than through a telephone (unless you know about modem hook-ups). If you have trouble with any of these procedures, ask an ARJCC employee for help. Experienced users working at the terminal site are also often willing to help beginners.

A. PRELIMINARIES

This section gives you some helpful definitions and describes our notation conventions.

—Definitions

User name: A user name is the name that allows the computer to recognize a particular user. Generally, the user's last name is taken as the user name. When you get an account, you will be told your user name.

Account name: Users are grouped into accounts. An account can correspond to a class, a staff, a group of researchers, or some other group. Accounts are named and you will be told what your account name is.

Password: Your password is the secret part of your user identification. The Computing Center usually takes your social security number as your password. You should change it to something else as soon as you can (section K of this manual).

File: A file is a unit of information storage. It can contain program statements, text, raw data, or any characters that the computer can encode. When information is put in a file, nothing else happens to it. For example, a program simply entered into a file is not processed (compiled, loaded, etc.) until specific commands, naming the file, are given later.

Page: A page is a part of a file. It holds about 2500 characters.

Directory: A directory is a group of files. Often a directory consists of a single user's files.

Most users have two directories. One is just used for logging in and the other holds the files the user creates. The name of the login directory is the same as the user name. The directory that holds a user's files is given a name that is a combination of the user name and the account name.

Examples:

a user name: SMITH
an account name: ECMP. 131
login directory: ⟨SMITH⟩
directory connected automatically: ⟨ECMP. 131. SMITH⟩

Terminal: A terminal is a device that allows users and the computer to communicate with one another. It consists of a keyboard and a means of displaying information, either paper or a video screen (cathode-ray tube or "CRT").

TOPS-20: The operating system of the DEC-20 is called TOPS-20. An operating system is a program that manages the sharing of the computer system's resources. You give it commands to do things like edit files, run programs, etc.

—Notation

[ret] This signifies that you are to press the RETURN or NEW LINE key once. We write [ret] and [esc] with adjacent spaces so that you can read the examples. Do not enter the spaces when you actually try the commands out.

[esc] Press the escape key ("ESC") or ALTMODE key.

@ The computer displays this character (called the "at sign" prompt) at the terminal when it is ready for a TOPS-20 command. You do not type it in for any of the examples given in this manual.

* This is the prompt symbol printed by the editor. Like the @, you do not type it in for any of the commands illustrated here.

—Examples

(. . .) In command illustrations, anything in parentheses is printed by the computer. You do not enter it.

When we use line numbers and file names in command illustrations, we are giving examples. Usually, you can use any specific line numbers or file names you want in commands.

B. LOGGING IN AND OUT

—To log in:

1. Type CTRL/C. (Hold the CTRL key down and press C.)
2. After you get the @ symbol (which prompts you for a command), type LOGIN and press [esc] (the escape key, labeled ALTMODE on some terminals.)
3. After you get the (USER) prompt, type your user name and press [esc].
4. After the (PASSWORD) prompt, type your password (which will not echo at the terminal). Then press [esc].
5. After the (ACCOUNT) prompt, type your account name and then press [ret] (the RETURN or NEW LINE key).

Example:

@LOGIN [esc] (USER) SMITH [esc] (PASSWORD) 301-52-1477 [esc]
                                      (ACCOUNT) MIDS. 403 [ret]

—To log out, type

@LOGOUT [ret]

—If you cannot log in, enter CTRL/C and then @HELLO [ret]

C. SOME SIMPLE COMMANDS

Try the following TOPS-20 commands.

@DAYTIME [ret]   Gives date and time.
@SYSTAT [ret]   Gives system status.

D. USING FILES

To create a file, you can give the TOPS-20 command to edit it. The editor auto-matically creates it when it can't find an existing file with the specified name. Beside the @ symbol, type EDIT and press the [esc] key. After (FILE), type a file name and then press the [ret] key.

@EDIT [esc] (FILE) TRIAL. TXT [ret]

The second part of the file name, .TXT, tells what kind of file TRIAL is; .TXT means "text". (File types are discussed in section G.) The editor says it is creating the file and then prints:

Input: TRIAL. TXT. 1
00100

Input (what you put into the file) is for version 1 of TRIAL. TXT. The first line is number 200. Type something on it and press [ret]. The system gives you line 200. Type some more and press [ret]. After the last line you want to enter, press [esc]. For Example:

00100 A line. [ret]
00200 Another line. [ret]
00300 A third line. [ret]
00400 [esc]

In this case, [esc] echoes as $. The system prints *. Type E and [ret] to end the edit session.

*E [ret]

File names: In general, file names (excluding the type specification) can be up to 39 characters long and can be composed of letters, numbers, and hyphens. If a file is to contain a program, it can be no longer than six characters and cannot include hyphens. Files to be used by many programs can be no longer than six characters, so it is a good idea to limit the length of all file names to six characters.

—To view the contents of TRIAL. TXT at the terminal:

   @TYPE [esc] (FILE) TRIAL. TXT [ret]

—To get a list of all your files:

   @DIRECTORY [ret]

Files you can see in your directory that have a Q as the first character of their type specifications are back-up files. You can delete them if you don't need them. (The system keeps only one back-up for most files.)

—To get rid of the file named TRIAL. TXT:

   @DELETE [esc] (FILES) TRIAL. TXT [ret]

DELETE marks a file for later deletion. When you log out, TOPS-20 will get rid of any files so marked. To get rid of such files immediately, follow the DELETE command with the EXPUNGE command:

   @EXPUNGE [ret]

—To print TRIAL. TXT on the lineprinter:

   @PRINT [esc] (FILES) TRIAL. TXT [ret]

The file(s) will be printed on the lineprinter closest to you.

E. EDITING A FILE

To make changes in TRIAL. TXT, give the edit command:

   @EDIT [esc] (FILE) TRIAL. TXT [ret]

The editor prints *, the prompt for edit commands.

—Print Command

*P200 [ret]   Prints line 200 at the terminal.
*P100:300 [ret]   Prints lines 100–300.
P ↑: [ret]   Prints all the lines on the current page. (P, up-arrow, colon, asterisk)

—Replace Command

Type R and the number of the line to be replaced. The editor presents the line number and you type in the new edition of the line.

    *R250 [ret]
    00250 A new edition of line 250. [ret]

—Insert Command

Type I and the number of the line that the insert is to follow. The editor selects and prints a line number for the insert and you type the line.

    *I400 [ret]
    00450 An inserted line. [ret]

If the last line in the file is named in the I command, pressing [ret] after the new line will get another new line. This continues until an [esc] is given.

You can insert more than one line between two existing lines. To do this, specify how inserted lines are to be numbered. For example, if you want lines inserted after line 700 and you want them numbered by intervals of 10, the command is:

    *I700,10 [ret]

Note that the 10 is included in the command after a comma. The editor will give you line 710, 720, 730, etc. as long as you press [ret] at the end of each new line. When you press [esc] or when there is no more room for insertions, the editor returns to the * prompt.

To insert a line at the beginning of a file:

*I [ret]   I, up-arrow [ret]

—Delete Command

*D675 [ret]   Deletes line 675.
*D100:550 [ret]   Deletes lines 100–550.

—Substitute Command

    *Smonkey[esc]ape[esc]14300[ret]   Changes every occurrence of "monkey" to "ape" on line 14300.

—Find Command

*Fmonkey[esc][ret]   Finds the line containing "monkey".

—Renumber Command

*N [ret]   Causes all lines in the file to be renumbered by increments of 100.

—Ending Editing

*E [ret]   Saves changes.
*EG [ret]   Ends, but makes no changes (quits).

Every time you edit a file, a new file with a new "generation number" is created to hold the updated information. (Old generations are deleted automatically, except for one backup.) You don't need to refer to the generation number when doing most things with a file.

F. CONTROL CHARACTERS AND DELETE

CTRL/U	Cancels partially typed line.
CTRL/W	Deletes most recent word entered.
CTRL/C CTRL/C	Aborts a running program.
CTRL/T	Gets information about session status.
CTRL/R	Reprints a line you are working on. (Use if display has become confusing.)
CTRL/S	Temporarily stops screen scrolling.
CTRL/G	Restarts screen scrolling.
DEL	Delete key. Deletes the last character you typed, similar to a backspace. (Uppercase on some terminals. Sometimes labeled RUBOUT.)

G. FILE TYPES

Program files have a type specification as part of their names.

TRIAL. FOR. 3	Generation 3 of a Fortran file.
JOB7. ALG. 2	Generation 2 of an Algol file.
ALICE CBL. 5	Generation 5 of a Cobol file.
PAPER. RNO. 9	Generation 9 of a Runoff file.

You can make up your own file types for your own reasons. For example:

SEMANTIC. MEMORY. 4
EPISODIC. MEMORY. 2

H. RUNNING PROGRAMS

Put your program in a file with the appropriate type. An example of a Fortran program:

@EDIT [esc] (FILE) TEST. FOR [ret]

```
00100       WRITE(5,101) [ret]
00200 101   FORMAT(' THIS IS A TEST. ') [ret]
00300       END [esc]
```

*E [ret]

To run the program:

@EXECUTE [esc] (FROM) TEST. FOR [ret]

If there are errors, correct the program file by editing it. Then give the execute command again.

I. COMMAND SHORTCUTS

You usually don't need to type in a whole command for TOPS-20 to understand what you want to do. When you have entered enough of a command to differentiate it from other commands (usually 2 to 4 letters), you can press [esc]. The computer will type out the rest of the command. This is called "command recognition". For example:

@ED [esc] IT (FILE)
System prints "IT
(FILE)".

You then enter the file name and press [ret] as usual. If what you type before [esc] is not enough to distinguish it from other commands, the terminal will beep and you must enter more letters of the command.

Pressing [esc], which usually gets the system to print a guide word, can be omitted in most cases. For example:

@EDIT PROGC. ALG [ret]

works the same as:

@EDIT [esc] (FILE) PROGC. ALG [ret]

Command recognition can also be used without pressing [esc]:

@ED PROGC. ALG [ret]

J. GETTING HELP

The HELP command gets descriptions of system features and the question mark (?) gets lists of commands that can be given at a certain point.

@HELP [ret]   Gets description of how the HELP command works.

To get TOPS-20 to print a description of how a system feature works, name the feature in the HELP command. For example:

@HELP MAIL [ret]

If you are uncertain about what you need to enter next in some situation, you can enter a "?" (without pressing [ret]. The system lists the things you can do at that point and then takes you back to where you were before you entered "?". For example:

@DE?   This gets a list of all commands that start with DE.

Then TOPS-20 prints "@DE" and waits for you to continue typing on that line.

K. CHANGING YOUR PASSWORD

To change your password:

1. Enter SET DIRECTORY PASSWORD and press [esc].
2. After TOPS-20 prints (OF DIRECTORY), type your login directory name (the same as your user name, which you use to log in) and press [ret]. Your login directory name must be typed inside angle brackets.
3. After the system asks for your old password, type it in and press [ret]. (Passwords will not print out when you type them.)
4. After the system asks for your new password, type it in and press [ret].
5. TOPS-20 will ask you to repeat your new password to make sure you didn't make a typing error. Press [ret] after you do so.

For example:

> @ DIRECTORY PASSWORD [esc] (OF DIRECTORY) ⟨SMITH⟩ [ret]
> Old password: 282-66-7714 [ret]
> New password: OATMEAL [ret]
> Retype new password: OATMEAL [ret]

It is a good idea to choose a password that is at least six characters long and is not easy for other people to guess. For example, do not use your spouse's name or the name of your cat.

L. MAIL

To send mail to another user, enter

> @MAIL [ret]

When the system prints "To:", list the users (by user name) that you want to send mail to and press [ret]. When the system prints "CC:", list the users who are to get copies and press [ret]. (If no copies are to be sent, just press [ret].) After "Subject:", give a title and press [ret]. Then type in your message and press [esc] when you are done. For example:

> @MAIL [ret]
> To: JONES [ret]
> CC: [ret]
> Subject: LAST HISTORY TEST [ret]
> Message (Terminate with ESC or CTRL/Z):
>
> SEDGEWICK, [ret]
>   WELL, HOW DID YOU DO ON IT? [ret]
>   -VIOLET [esc]

To read your own mail, enter

> @RDMAIL [ret]

When the system asks for a date and time, you can just press [ret] again and you will be shown any messages you have not read previously.

Appendix D
Alternate Fuels
Demonstration Program:
Excerpts on Solar Energy

8. Wind Energy

Accelerated wind energy development could produce energy that could be substituted for about 20 percent of the output of the program at a 1 million barrel a day level in 1985; it could not be an alternative to the entire program at either a 350,000 or a 1 million barrel a day level in 1985.

9. Hydroelectric Energy

Due to the long lead time (approximately 20 years) required to bring hydroelectric power on-line, increased hydroelectric power could not be a feasible alternative. However, the Department of Defense (Corps of Engineers) together with other responsible agencies will report to the President on the potential for installation of additional hydroelectric generating capacity at existing dams throughout the country.

10. Tidal Energy

Due to technological problems and the long lead time that would be required to bring tidal energy on-line, tidal energy could not be a feasible alternative.

11. Tar Sands

Accelerated tar sands development could produce energy that could be substituted for about 20 percent of the output of the program at a 1 million barrel a day level in 1985; it could not be an alternative to the entire program at either a 350,000 or 1 million barrel a day level in 1985.

12. Increased Domestic Oil and Gas Production

Increased domestic oil and gas production could be an alternative to the program at a 1 million barrel a day level in 1985. The primary constraint to increased oil and gas production is the economics of the increased production. Potential impacts would include industrialization of previously remote areas, loss of terrestrial and aquatic habitats, and oil spills.

13. Hydrogen Economy

Hydrogen is not a primary source of energy and would have to be produced from such primary sources as coal, solar energy, nuclear energy, etc. The Hydrogen Economy concept is an alternative to the program to the extent that the coal used to produce synthetic fuels could be used to produce hydrogen instead. Potential impacts would be the same as those discussed under the Direct Utilization of Coal Alternative.

14. Energy from Solid Wastes

Accelerated production of energy from solid wastes, either by incineration or anaerobic digestion, could be substituted for about 40 percent of the output of the program at a 1 million barrel a day level in 1985; it could be an alternative to the energy production and security objectives of the program at a 350,000 barrel a day level in 1985, but could not be an alternative at a 1 million barrel a day level in 1985. Through the Resource Conservation Recovery Act, the Federal Government will continue to help state and local governments reduce barriers to the production of energy from solid wastes.

G. *Solar Energy*

1. Background

In the past, solar energy provided a major share of the energy consumed by pre-industrial and early industrial societies. For the most part, the sun's energy was used indirectly, primarily in the form of wind and firewood. However, burning glasses that were used to light fires have been found in the ancient ruins of Assyrian cities dating back to the 7th Century B.C., and it is claimed that Archimedes used such devices to set fire to Roman ships attacking Syracuse in 212 B.C. One of the earliest uses of solar energy was to distill salt from brackish water in arid regions.

In more recent times, more direct use of solar energy has been accomplished. Thermal conversion to mechanical energy was demonstrated in the latter part of the 19th century. A workable solar-powered water pump was displayed by Mouchot at the Paris World Fair in 1878. Frank Shuman, an American engineer, built a 100-hp solar-powered steam engine in Egypt in 1912 (Wolf, 1973). However, all such attempts at a practicable thermal conversion system have been less than satisfactory, primarily because collectors have not been developed that would collect solar energy efficiently and, at the same time, provide high temperatures necessary for good efficiency of a heat engine.

Most contemporary uses of solar energy provide thermal energy for buildings. Solar water heaters are manufactured and used in several countries including the United States; they were once common in Florida, but their use has diminished because of the availability of natural gas. Space heating by use of solar energy has been demonstrated in about 20 experimental buildings (NSF-NASA Solar Energy Panel, 1972). Under National Science Foundation (NSF) sponsorship, four schools in different parts of the country have recently been equipped with solar collectors to provide portions of the heating loads (Hauer, 1974). Some experimental work on solar-powered air conditioning by use of absorption refrigeration has been conducted, and an experimental house that uses solar energy for both heating and cooling has been constructed at Colorado State University (Hauer, 1974). This house is to be used as a solar energy laboratory for government-sponsored research. Solar energy systems for residential and commercial buildings that combine water heating, space heating, and air conditioning are considered to have the most promise because solar collectors and energy storage units are common to all three functions.

Photovoltaic devices, that convert the sun's energy directly to electricity, are being used extensively on space vehicles as primary power supplies. Solar arrays, consisting of these same types of photovoltaic devices, have been used with battery storage to supply power for telephone exchanges in the southern U.S.

2. Technological Status

Solar energy conversion technologies are presently in an expanding phase of research and development, and are not now contributing significantly to the Nation's energy supply.

There are several solar energy conversion systems that are currently the subject of research programs designed to determine their potential energy contribution and to accelerate their ultimate utilization. These systems fall into the following categories:

- solar heating and cooling
- solar thermal conversion
- process heat
- photovoltaic electric power
- ocean thermal conversion
- wind energy
- bioconversion.

The conditions in the Earth's crust leading to the presence of some geothermal phenomena are also those conditions producing faults and earthquakes. As a result, in some instances geothermal and seismic phenomena are geographically inseparable. Experience in petroleum production indicates that marked changes in reservoir pore pressure, whether due to the withdrawal of fluids or to injection, may in certain types of reservoirs (especially in faulted or fractured rocks) result in instability leading to micro-earthquake activity. Such instability due to production alone has been documented in the Wilmington Oil Field, California (Poland and Davis, 1969). Instability due to injection was documented

at the Baldwin Hills Oil Field, California (Hamilton and Meehan, 1971), and at
the Rangely Oil Field, Colorado (Healy et al., 1970), and in connection with
injection of waste waters at the Rocky Mountain Arsenal, Colorado (Healy et al.,
1968). Both withdrawal and reinjection may occur in geothermal fields.

However, reinjection may not present a seismic problem in many geothermal
fields. Those instances in which there has been instability due to reinjection
have involved reinjection at pressures exceeding hydrostatic. In such instances,
reinjection could open and lubricate preexisting fractures and zones of weakness
or extend the fracture pattern, causing increased seismic activity and perhaps
structural damage. However, a number of potential reservoirs (e.g., dry steam
systems) are at subnormal pressures (less than hydrostatic). If the return of
fluids merely maintains preexisting pressures in such reservoirs, it should not
cause the increasing seismicity noted in other conditions (Bowen, 1974).
However, if pressures were to significantly increase, there could be seismic
activity. There is no evidence that geothermal production has increased the
seismicity of an area.

In addition, whenever fluids are extracted from a groundwater reservoir, there
is a potential for land subsidence to occur. Some geothermal reservoirs may be
especially susceptible to this phenomenon because of the high mass removals
required for production. Subsidence has occurred at geothermal fields in
Wairakei, New Zealand and Cerro Prieto, Mexico after the extraction of
geothermal water without reinjection.

Because of the geologic circumstances under which dry-steam fields develop,
subsidence is not expected to occur in them. However, hot-water fields could be
subject to subsidence, especially if pressures are not maintained by fluid return
(Bowen, 1974).

b. *Potential Modification to the Biological Environment*

Impacts from construction and plant operation would be similar to those
discussed in Chapters IV and V and are not repeated here. Principal impacts
would result fom land disturbance and activities in remote and undisturbed
areas.

c. *Potential Modifications to the Socioeconomic Environment*

Impacts to the socioeconomic environment would be similar to those discussed
in Chapter IV and V and are not repeated here. The potential impact would
depend upon both the relative size of any population increase associated with
the geothermal development and the rate of such an increase.

a. *Solar Heating and Cooling Systems*

Solar heating and cooling systems utilize solar radiation for the comfort heating
and cooling of residential, commercial and industrial building space; the
provision of hot water; and certain technologically-related agricultural

applications such as greenhouse heating and crop drying. These energy needs now consume between a quarter and a third of all the energy used in the United States (Interagency Task Force on Solar Energy, 1974).

In typical heating and cooling systems, solar energy, incident on a collector surface, is converted to thermal energy in a working fluid—commonly water or air. This working fluid then delivers the heat energy to the conditioned building space, to thermal energy storage equipment, to heat-operated refrigeration equipment, or to equipment to heat water for use in the building.

The technical feasibilities of solar space heating and hot water heating have been adequately demonstrated. Solar cooling devices of several types are in the advanced development or demonstration stage (Interagency Task Force on Solar Energy, 1974; The MITRE Corporation, 1973). No technological breakthroughs are required to construct and operate systems of these types.

The principal factors presently limiting the broad application of solar space conditioning equipment are lack of low-cost collectors and the present state-of-the-art of heat driven refrigeration systems. Initial costs of solar space conditioning systems are presently higher than conventional systems although life-cycle costs of solar systems are expected to be significantly lower. Lower initial costs of solar energy systems should be brought about through engineering development and large-scale production.

b. *Solar Thermal Conversion Systems*

Solar thermal conversion (STC) systems collect solar radiation and convert it to thermal energy for use as the heat input to systems for generating electric power. The technologies required for STC systems include (a) the collector system, (b) a system for transferring the collected energy to a working fluid or to a storage system, and (c) a system for converting heat to electrical energy, probably using turbine-generator components of a conventional electric power plant.

The plant factor of the collectors is expected to be only about 25 percent, even in regions of high insolation such as the southwestern parts of the U.S. Researchers have pointed out that for solar energy plants to provide firm power, several days of energy storage will be required (Meinel, 1974; The MITRE Corporation, 1973). Solar electric systems with limited energy storage would appear to need full capacity backup from conventional plants. Thus, most proposed solar thermal-conversion plants would not appear to be stand-alone systems, but would provide supplemental power. Except for the development of storage systems, there appear to be no fundamental limitations which restrict the development of STC technology. The introduction of STC systems would be primarily dependent upon reduction in capital costs, optimization of unit size, construction time requirements, and the economics of competing energy sources. Applications of STC systems would be most promising in regions of high insolation. It is expected that a major application will be for intermediate load service in the southwestern region.

c. *Process Heat Systems*

Process heat systems are similar to STC systems except that the thermal energy is used to supply the process heat required for industrial and other applications rather than to generate electricity.

Process heat systems are of three types: low temperature systems which could be used for fog disposal, melting of snow or ice, evaporation and distillation of water, food processing, etc.; medium temperature systems which could be used for general industrial processes requiring heat or steam; and high temperature systems which could be used for special industrial heat processes.

These systems appear technically feasible and, as the price of fossil fuels rises, show promise of being more and more economically competitive for a number of types of applications, e.g., industrial processes, agricultural processes, aquaculture, saline water distillation, etc. The main drawbacks of process heat systems are the extensive use of land that would be required by the solar energy collectors, the maintenance of evacuated components and moving parts of the collectors that may be necessary for high temperature applications, the need for large thermal storage capabilities for many applications, and the relative inflexibility of these systems, in all applications, compared to fossil-fuel process heat systems.

Process heat systems need further research and development and proof-of-concept experiments to prove their economic viability for those applications best suited for use of solar systems (The MITRE Corporation, 1973).

d. *Photovoltaic Conversion Systems*

Photovoltaic conversion systems use semi-conductor devices to convert sunlight directly into direct current electricity or fuel gas at efficiencies ranging from 5% to 20%. Efficiency depends on the materials employed and the design of the solar cells and arrays.

Solar cells which convert sunlight directly to electricity have been widely used in the past for photoelectric cells and other devices. In recent years they have found applications as primary sources of power for a number of types of spacecraft. Photovoltaic systems appear well adapted for use in producing not only electricity, but also hydrogen, which can be used either as a fuel gas or as an energy storage medium.

For space applications, the main objectives have been to develop highly efficient and reliable, long-life, light-weight arrays of solar cells, with cost being a secondary consideration. In terrestrial applications, on the other hand, the main objectives would be to develop highly efficient and reliable, long-life, low-cost arrays of solar cells, with weight being a secondary consideration.

Principal drawbacks of photovoltaic systems are the extensive land areas required by their solar energy collectors and the low plant factors for the collectors, i.e., only about 25%, even in regions of high isolation, such as the southwestern parts of the U.S. Also solar cells are still relatively expensive and low in efficiency.

Photovoltaic systems will require an extended set of proof-of-concept experiments, and continuing research and development (until perhaps 1990) to develop low-cost methods of producing pure silicon and other possible solar cell materials, as well as to develop methods of low-cost mass-production of highly efficient solar cells made from these materials (The MITRE Corporation, 1973). A technological breakthrough may be needed to meet economic objectives.

e. *Ocean Thermal Conversion Systems*

Ocean thermal energy conversion (OTEC) power plants would extract heat from ocean surface waters, utilize this heat to drive a turbine, and reject heat to the heat sink provided by the cold ocean water at lower depths. Energy would be obtained by exchanging heat between the ocean water and a working fluid such as ammonia or propane, operating in a "closed-cycle" similar to a refrigeration cycle. Thus, a total system operating at sea would include a semi-submersible hull incorporating power-pack modules. Each module would contain an evaporator, turbine, condenser, and pumps to circulate the working fluid and the warm and cold water.

Use of the ocean thermal resource is projected to commence in 1985 and to exhibit rapid growth in the 1990's. Commercialization of OTEC technology is dependent upon the solution of a number of engineering and institutional problems that would permit the development of cost-effective power plants (Interagency Task Force on Solar Energy, 1974).

f. *Wind Energy System*

The generation of electricity from wind has been extensively developed in recent years in a number of countries. The use of wind energy is an indirect use of solar energy and is discussed as a separate alternative to the Synthetic Fuels Commercialization Program in a later section of this chapter.

g. *Synfuels from Biomass*

The overall objective of the Fuels from Biomass (FFB) program is to develop the capability for converting renewable biomass resources into clean fuels, petrochemical substitutes, and other energy-intensive products that can supplement similar products made from conventional fossil fuels. The program offers a significant potential for reducing this country's dependence on fossil fuels in the mid-to-long term, through the conversion of a renewable energy source to liquid and gaseous fuels, electric energy and petrochemical substitutes. Synthetic fuels from biomass resources would not be expected to make a significant energy contribution before 1985–1990.

The energy-intensive products which can be produced from biomass are, in order of fuels from Biomass program priority:

- *Liquid Fuels*—The most important are methanol (wood alcohol) and ethanol (grain alcohol). These can be mixed with gasoline to extend supplies of motor fuel, or used as peaking fuels to generate electric power.

- *Gaseous Fuels*—These include medium-BTU gas for on-site application, upgraded synthetic natural gas (SNG) fed into natural gas pipelines, and hydrogen.
- *Petrochemical Substitutes and Other Energy-Intensive Products*—These include ammonia, higher alcohols, Ketones, turpentine, and resins.
- *Heat from Direct Combustion*—This is used to produce process heat and to generate pressurized steam and electric power.

Conversion processes for biomass are in various stages of research and development, but few will be commercial by 1985. Direct combustion of biomass (wood for example) can compete in several regions of the Nation where biomass feedstock supplies at reasonable costs can be assured. In addition, anaerobic digestion of manures from environmental feedlots is very close to commercialization. Other conversion technologies under development include gasification, liquefaction, and fermentation.

The primary biomass resources being considered are: forest and agricultural residues, terrestrial energy farms to grow biomass for energy, and aquatic energy farms. In the near term, the available resource will be residues; research and development on the growth and harvest of promising species is underway.

The utilization of organic waste materials is another indirect use of solar energy and is discussed as a separate alternative to the Synthetic Fuels Commercialization Program in a later section of this chapter.

3. Potential Energy Contribution

a. *Resource Availability*

The solar input to the earth averages 130 watts per square foot for a total of 1.78 $\times$ 10^{17} watts or 5.3×10^{21} Btu per year, of which about 3.6×10^{21} Btu per year are retained within the earth's atmosphere and surface and roughly 2.4×10^{21} Btu per year are absorbed by the surface (Gustavson, 1971). Each year the land area of the U.S. intercepts solar energy equivalent to approximately 600 times the current domestic energy requirement (U.S. Department of the Interior, 1973). At an average energy conversion efficiency of five percent of the incoming solar energy flux, less than four percent of the U.S. continental land mass could supply 100 percent of the Nation's current energy needs (Interagency Task Force on Solar Energy, 1974).

Four characteristics of the solar energy resource base are of particular interest:

- it is a diffuse, low-intensity source of energy
- the energy is spread over various frequencies (i.e., distributed over the various wave lengths of light
- its intensity is continuously variable during the daylight hours, is zero at night, and is subject to weather and seasonal variations
- its availability differs widely among geographic areas.

b. *Constraints*

The number and range of potential solar applications are extensive but the present state-of-the-art is such that energy collection efficiencies are low, and the requirements for energy storage resulting from the intermittent nature of the source result in costs that are prohibitive for general use. With the exception of bioconversion and ocean thermal conversion, all solar energy systems require some storage capability. Conventional backup systems must be utilized to the extent storage is inadequate during periods of no insolation or wind.

A typical 1,000 MW(e) power plant operating in a 1,400 Btu/day solar climate would, with present technology, require 39 square miles of collector surface (assuming a 30 percent efficiency in converting solar energy to process heat and a 5 percent efficiency of conversion to electrical energy) (The MITRE Corporation, 1973).

Table XII-32 presents a projection of the price of electricity produced by various solar energy systems compared to the price of electricity produced by conventional systems. Other sources have also indicated the high cost of solar energy compared to conventional energy (Oak Ridge National Laboratory, in press; Cherry, 1972; Löf, 1973).

The use of solar energy is now uneconomical, except for solar heating and cooling applications. The outlook appears to be that solar energy has limited potential as an economical, major source of electricity since 1985. The solar application that potentially could play a significant energy role in the mid-term is thermal energy for the heating and cooling of buildings.

c. *Solar Energy Potential in 1985*

Three projections of future solar energy contributions for producing electricity, thermal energy and high energy fuels are summarized in Table XII-33. All

Table XII-32   Projected Economic Viability Ratio of Solar Energy to Conventional Energy

Solar Energy System	Ratio of Solar Energy Costs to Conventional Energy Costs		
	1985	2000	2020
Solar Heating and Cooling	1:1 to 1.5:1	1:1 to 1.25:1	−1:1
Process Heat	1.5:1 to 2:1	1:1 to 1.5:1	−1:1
Solar Thermal Conversion	2.5:1 to 3:1	2:1 to 2.5:1	−1:1 to 2:1
Ocean Thermal Conversion	1.5:1 to 2:1	−1:1	−1:1
Photovoltaic Conversion	6:1 to >10:1	1:1 to 4:1	−1:1
Wind Energy	−1:1	−1:1	−1:1
Bioconversion	−1:1	−1:1	−1:1

Source: The Mitre Corporation, *Systems Analysis of Solar Energy Programs*, MTR-6513, McLean, Virginia, December 1973 (with modifications).

Table XII-33 Projections of Potential Solar Energy Applications

	Market Penetration by Year Given(a)					
	2000			2020		
Application	NSF-NASA(b)	Subpanel IX(c)	MITRE(d)	NSF-NASA(b)	Subpanel IX(c)	Mitre(d)
Thermal Energy for Buildings	10%	30%	18%	35%	50%	31%
Electric Energy						
Solar Thermal Conversion	1%	40,000(e) MW(e) (peak)	12%–29%	5%	30%(g)	29%
Photovoltaic Conversion	1%	100,000(e) MW(e) (peak)	12%–29%	10%	10%	29%
Combustion of Organic Material	1%			10%		
Total Electric				20%(h)		
High Energy Fuels						
Gaseous	10%(g)	50%(f)		30%(i)		
Liquids	1%(g)			10%(i)		
Other						
Organic Fuels(i,j)				(7%)		(10%)
Process Energy (j)				(1%)		(10%)

(a) Figures show the percent of each energy market captured by solar energy unless otherwise specified.

(b) NSF-NASA Solar Energy Panel, "An Assessment of Solar Energy as a National Resource," Washington, D.C., December 1972.

(c) Subpanel IX, "Solar Energy Program," A. J. Eggers, Jr., Subpanel Chairman, National Science Foundation, Washington, D.C., November 13, 1973.

(d) MITRE Corporation, "Systems Analysis of Solar Energy Programs," MTR-6513, McLean, Virginia, December 1973. Projections are in terms of national energy consumption. Table values were computed using data on building thermal energy from Report AET-8, year 2000 electrical production from Dupree and West, and year 2020 electrical production from Report AET-8.

(e) Subpanel IX report gives various values for photovoltaic market projections including 1%, 2%, 5%, and 7% of electrical capacity by the year 2000. The 100,000-MW(e) (peak) figure was obtained from their year-by-year tabulation of installed photovoltaic capacity. This peak capacity would correspond to about 2% of projected electrical generating capacity in the year 2000. Similarly, solar thermal conversion peak capacity corresponds to less than 1%.

(f) Six percent from organic wastes and 44% from managed production of photo-synthetic materials. Time period for achievement of given percentages unspecified in Subpanel IX report.

(g) Subpanel IX indicates 30% if the ultimate potential of solar thermal conversion is achieved. Time period unspecified.

(h) Includes energy generated by wind and ocean thermal gradients.

(i) Includes high energy fuels from organic wastes as well as managed production of photosynthetic materials.

(j) Parenthesized figures are percentages of national energy consumption.

projections are based on the assumption that a substantial research program would be undertaken for each of the solar energy systems and that these programs would result in economically viable alternatives to conventional energy sources. All three projections agree that the most significant near-term impact of solar energy would be in the heating and cooling of new buildings.

Table XII-34 shows an estimate of the maximum potential solar energy contribution in 1985 for a business-as-usual implementation scenario and an accelerated scenario (see Interagency Task Force on Solar Energy (1974) for a discussion of the scenarios). The entries in the table represent an estimate of the upper bound on the implementation of these solar energy technologies. The assumptions behind this implementation include the projected success of the recommended research program in solving the economic, technical, sociological, and institutional problems of solar energy systems; no limitations on the availability of capital needed to implement the required solar energy systems; reasonable assumptions on U.S. production capacities and materials availability; and achievement of the estimated market penetration for these types of systems in the face of competition from other energy sources, but not including competing solar energy systems. In actual practice, any of the above types of constraints could limit the projected contribution of some or all of these solar energy technologies, thereby decreasing the individual estimates for each technology shown in Table XII-34.

It would be very optimistic to expect that all of these technologies would achieve the maximum levels shown in the table. Rather, some mix of these technologies would probably be implemented, thereby resulting in a lower total annual contribution of solar energy systems than would be indicated by the sum total of all of these estimated upper-bound contributions.

A maximum potential exists for solar energy to replace the equivalent of an additional 315,000 barrels a day of oil in 1985 if the solar energy program were to be accelerated. (This assumes that the business-as-usual replacement of 558,000 barrels per day of oil equivalent would occur regardless of the proposed Program's existence.) Thus, solar energy acceleration cannot be viewed as an

Table XII-34 Estimates of Maximum Potential Solar Energy Contribution in 1985 (Barrels a Day Oil Equivalent)

Solar Energy System	Business-as-usual	Accelerated Program
Solar Heating and Cooling	500,000	800,000
Solar Thermal Conversion & Process Heat	3,000	3,000
Ocean Thermal Conversion	50,000	50,000
Photovoltaic Conversion	5,000	20,000
Total	558,000	873,000

Source: Interagency Task Force on Solar Energy, *Project Independence Blueprint Final Task Force Report,* Washington, D.C., November 1974. Based on oil at $11 a barrel and a 30 percent efficiency in converting oil to electricity.

alternative to the entire Program; however, if accelerated, solar energy could be substituted for up to about 31.5 percent of the output of the Program at a 1 million barrel a day level in 1985 and up to 90 percent of the output at a 350,000 barrel a day level.

4. Environmental Impacts

Nearly the entire solar energy contribution (300,000 out of the 315,000 barrels a day) would be in the form of solar heating and cooling systems. These systems would have practically no environmental impact since they would normally be integral parts of the buildings they serve (in the case of large buildings, the units would likely be located on the surrounding land). Very little water would be required for space heating and that used would be recycled. Normal amounts of water would be used for the hot water supply. There would be no significant change in the heat balance of the environment from the use of this type of system. The only impacts from the system itself may be in the esthetics of the system and the noise it produces as well as the danger of an occasional leak. (The MITRE Corporation, 1973). Construction of facilities to fabricate the units and transportation of the units would cause the same types of impacts as discussed in Chapter V.

The remaining 15,000 barrels a day oil equivalent solar energy contribution would be from photovoltaic systems. The major impact from such a system would be from the land area required for the collector arrays. Approximately 15 square miles of land would be required for the collectors (The MITRE Corporation, 1973).

A small amount of water would be required for the collector array for cooling the solar cells. Water would be required for the electrolysis units if hydrogen is produced for energy storage, or for use as a fuel gas of industrial feedstock.

No external noise or odors would be generated by the photovoltaic system. However, the collector structures, the transmission lines, and the buildings housing the inverters and the electrolysis units, if utilized, would change the nature of the surrounding area.

Solar collectors could affect the heat balance of the siting area. The collectors would alter the albedo in the site area. The area below the collectors would be shaded while the area above them would likely receive an increase in solar energy as a result of the reflection of the solar energy by the collectors. The net effect would be a withdrawal of heat from the site area to the load points of the system. The temperature differential above and below the collectors would to some degree affect wind conditions in the site area. In addition, surface wind velocity would likely be dampened to a small degree by the impediment of the collectors. The precise effects of all these changes cannot be predicted at the current time (Energy Research and Development Administration, Division of Biomedical and Environmental Research, 1975).

Collectors would shade a significant portion of a site and could prohibit the reestablishment of original plant communities following construction. Other species not affected by construction may be affected by reductions in sunlight. Moderation of the extreme surface micrometeorological conditions that are typical of desert environments may promote the growth of plants suited to more moderate conditions and may increase primary productivity. Microclimatic changes emanating at a project site may also affect vegetation in adjacent areas. The washing of heliostats may provide supplemental water to the site area and may also encourage the growth of vegetation.

Destruction of habitat would affect the fauna of an area as discussed in Chapter VI. However, the amelioration of extreme micrometeorological conditions in desert areas and any resultant increase in vegetation could benefit some populations of invertebrates and small vertebrates. Changes in microclimate may also affect animal habitats in adjacent areas. The concentrated radiant flux directed from heliostats to a receiving tower could create a hazard to birds and insects within the area of the project site, and the bright reflecting surfaces of the collectors may interfere with movements of bird populations in the area. Local impacts of the facility on thermal currents may also affect the distribution of raptors and other soaring birds. Security fences, if they are erected, would interfere with the movement of terrestrial animals. Solvents used to clean collectors could also affect animal habitats on and near the site.

Glare from collectors would represent a potential hazard to very low flying aircraft. However, because each reflector would be focused separately, their effect is expected to be similar to flying across the choppy water of a lake. Some turbulence could also be created by any heat plume rising above the reflectors. The combined effect of these two impacts is expected to be slight, but could necessitate modifications to flight paths of very low flying aircraft. Changes in flight paths could result in increased noise and air pollution in areas over which the modified flight paths crossed.

Index